T0220496

FROM THE TOP OF THE MOUNTAIN

FROM THE TOP OF THE MOUNTAIN

Rosolino Buccheri

JENNY STANFORD
PUBLISHING

Published by

Jenny Stanford Publishing Pte. Ltd.
101 Thomson Road
#06-01, United Square
Singapore 307591

Email: editorial@jennystanford.com
Web: www.jennystanford.com

British Library Cataloguing-in-Publication Data
A catalogue record for this book is available from the British Library.

From the Top of the Mountain

ISBN 978-981-5129-22-9 (Paperback)
ISBN 978-1-003-50829-8 (eBook)

To my wife, Riri
for 50 years together

Et ainsi, les actions de la vie ne souffrant souvent aucun délai, c'est une vérité très certaine que, lorsqu'il n'est pas en notre pouvoir de discerner les plus vraies opinions, nous devons suivre les plus probables; et même, qu'encore que nous ne remarquions point davantage de probabilité aux unes qu'aux autres, nous devons néanmoins nous déterminer à quelques-unes, et les considérer après, non plus comme douteuses, en tant qu'elles se rapportent à la pratique, mais comme très vraies et très certaines, à cause que la raison qui nous y a fait déterminer se trouve telle.

DISCOURS DE LA MÉTHODE

Contents

Foreword

Rosolino Buccheri, an outstanding Italian astrophysicist, the author of the most recent, highly commendable essay,* continues herewith his dialog with readership and aims here at sketching the principles of research work methodology.

First and foremost, to properly attack this interesting and important problem, Prof. Buccheri goes back to the newest history of natural sciences, in stressing his very intent by a citation from René Descartes' *Discours de la Méthode* in the epigraph, and inviting us thereafter to cast a closer look at the problems of

(a) *Second Law of Thermodynamics, according to him, just to "help us stop any cold and static view concerning the drastic and negative perspectives of the Second Principle of Thermodynamics with its view of a Universe without any scope together with the end of any kind of life in a far future," and, consequently,*

(b) *Entropy notion, as to which he deals with the poser:*
   Entropy: a law or just a principle?

With this in mind, first and foremost, it is appropriate to analyze the Cartesian idea embodied in the epigraph (below comes its somewhat paraphrased English translation):

(a) *Our everyday actions do often permit no delay;*

(b) *It is a very certain truth that, when it is not in our power to discern plausible opinions, we ought to choose the most probable alternative and follow it;*

(c) *Although we sometimes do not notice any advantage of probability, we must nonetheless determine ourselves to the latter, and consider them no longer as doubtful, in so far as they relate to practice, but as very true and very certain, because the reason which made us determine there is such.*

---

*Rosolino Buccheri, *Myth, Chaos, and Certainty: Notes on Cosmos, Life, and Knowledge*; Jenny Stanford Publishing Pte. Ltd., Singapore, 2021.

Methodologically, this boils down to a kind of '*operational positivism*,' whereby—from the Cartesian past up to nowadays—we stubbornly tend to employ Probability Theory and mathematical statistics as soon as we encounter something unintelligible—but are **in a hurry** to deliver plausible answers to posers we tackle—and (*most importantly*!) **practicable recipes we do need to succeed in our field as soon as possible**.

Remarkably, it is just the above methodological trend that molds the keynote of the '*Modern views: Quantum, Prigogine, and contemporary science*' so profitably different from the '*Old, Classical*' legacy apparently eliminable, according to Prof. Buccheri.

Having just said this, the present writer recommends that young colleagues attentively read the deliberations to follow and do not hesitate to cast a look at a concise epilogue at the end of the story at hand.

Dr Evgeni B. Starikov

Graduate School of System Informatics
Kobe University
1-1 Rokkodai, Nada, Kobe 657-8501, Japan

Chemistry and Chemical Engineering
Chalmers University of Technology
Göteborg, Sweden

Introduction

This work tries to show that life—both animal and human—
is the direct consequence of the spontaneous evolution and
complexification of the inanimate matter under the action of
all natural forces and, as such, it is impossible that it has
developed only on our Earth, a very small grain of the immense
universe where we may easily imagine the existence and
the validity of the same physical laws. A universe that must
be full of life in any part of it, even if, due to the enormous
distances between different points of our galaxy or, even more,
between different galaxies, we will never know it.

My viewpoint avoids, as much as possible, falling into the
man's presumption to be the only intelligent being within the
immense universe, anyway destined to perish together with
it. Hard-to-die arrogance, this, perhaps derived by the many
constraints, both theoretical and practical, put in place by the
numerous beliefs and religions in the world, that due to the
impossibility to look into details within the Cosmos, are bound
to the idea that life is special and unique on our Earth, and
that it is bound to disappear together with Earth itself and
with anything else included here.

The average human life duration is less than one century,
and our knowledge is bound to the experiences done in such
a tiny space and time durations in which we live, space and
time *interval* enormously smaller compared to the one in which

From the Top of the Mountain
Rosolino Buccheri
Copyright © 2025 Jenny Stanford Publishing Pte. Ltd.
ISBN 978-981-5129-22-9 (Paperback), 978-1-003-50829-8 (eBook)
www.jennystanford.com

the entire universe evolves. Consequently, both the pleasure of our achievements and the grief of our mishappenings depend on what we may happen to experience in such a tiny period of our biologic life and on our ability to comprehend and control our ambient, limited to this extremely small space-temporal interval that peremptorily binds us.

Within such limited space and duration, and even by living a variously dynamic life, the status of the world appears to us sufficiently stable. Its physical structure appears to us as having remained grossly unchanged for millennia, the landscapes do not change too much with time—if it happens it is mostly due to man's activity—and the climate, sometimes very variable or even devastating, appears to let us leave a 'normal' life forever. Pacific or conflictual relations of any kind seem to have remained approximately stable under changing ideologies or traditions, while our single lives take turns with each other through successive generations, with continuous enrichments of knowledge that very slowly modify the mental models of peoples.

We know that our universe evolves at a much ampler space-temporal scale with respect to Earth: along millions of years, stars and planetary systems are born, grow up and die destroying each other and then self-renovate them through continuous collisions provoked by natural forces. In such enormous intervals of time, new generations of stars substitute those already exhausted and extinct, continuously modifying the general scenario. We foresee that within 4–5 billion years, our Sun, become a *red giant* after having run out all its nuclear fuel, will incorporate and destroy the entire planetary system together with human life on Earth. Such a disaster, anyway, could happen even before, possibly by means of an unexpected planetary collision or other space calamities, always possible. In these cases (to which we may include the possible human self-harm, sometimes prevailing over the reason), all what our

Earth will have produced until then, human life included, will cease to exist.

Despite such expected negative future, it may not be wrong to hypothesize that in the meantime, with some billion years hypothetically available, our technology could bring new and unexpected surprises, allowing us to safely move toward other planets situated in younger stellar systems. In such a case, the same physical structure of the human being, which originated on Earth and used to live there, will have to radically change in order to adapt itself to the new existence conditions, when traveling for centuries between unknown planet systems within the limited ambient of a spaceship in absence of gravity, or even in presence of the different atmospheres and gravities of other worlds. About all that, we do not have any need to worry, because today we cannot have full conscience about the enormous difference existing between the time of our single lives and the long evolution time of the full universe.

ᘓᖇ°ᕼᘒ

Concerning the physical origin and the end of our universe, we have good hypotheses, not definite certainties. Present scientific data tell us that the universe is presently going along the way of an endless expansion caused by a very violent *Big Bang*. A universe that, starting from very simple elements—molecules of hydrogen and helium—has been filled up, little by little, with a lot of new structures, becoming along the way always more complex, until the advent—certainly on Earth—of humans, those very complex living and thinking structures that we are and that seem to go toward a continuous increase in time of their cerebral abilities.

Independent of such a continuous increase of the livings' complexity, however, the present theorical speculations, nowadays largely supported all over the world, predict an eons-lasting expansion of the universe due to the ongoing escape velocity of matter for which the space among all single structures will continuously increase, causing a decrease of the gravity attraction and then of the aggregation power of matter. A situation that, in a far future, is foreseen to bring the entire universe toward an end, called 'thermic dead,' a very simple final state where all galaxies are destined to collapse toward several *Black Holes* stepping infinitely away from each other. A state in which human life is excluded and where the only possible complexity is given by the *Black Holes* themselves that, according to Stephen Hawking, can emit energy by means of quantum phenomena, thus causing the birth of new universes with new and unknown physical characteristics, not necessarily adapt to human life. In summary, a mix of uncertainties tamed by presumed certainties; a mix helped by the always ampler scientific knowledge available together with specific visions of life crystallized along time. Knowledge and *a priori* beliefs always twisted together!

Anyway, beyond any kind of self-assurance discussed in literature, our possibility to understand the world in all its details cannot avoid depending on the great difference between our small dimensions and the large ones of Earth, or especially those enormous ones of the entire universe, whose immensity escapes at all from our mental control. A vast and complex reality in which we are merged, but of which we are only able to see and evaluate an extremely small part of it, with only imaginative possibilities to understand the greatness in which we are involved. A condition, this, which implies important physical and knowledge limitations, impossible to fill up with the limited experience we may store on Earth.

Such a condition is very natural for us. Due to our physical and cerebral limitations, we cannot be able to keep in mind consciously and continuously all the enormous quantity of information of so vast and contrasting nature that floods in our brain every single moment of our life. So, it happens that we can log into fine details only in a very limited way and, in any case, only in some specific conditions of our life. Limitations that are not certainly equal for all of us, but from which nobody can escape at all, thus continuously forcing us to decide whether to give the most of our attention to what concerns the everyday life so to manage at our best the necessary details, or to spend a consistent amount of our energy for looking things from the height, in order to better understand our relation with the external realities, even risking to neglect important details for the everyday life. This is the way done by the specialists of any sector who continuously risk falling into error when approaching more general topics. In any case, absolute certainties are never possible, whatever one may ever believe.

Together with the need for such a choice, another problem is in front of us, of even greater importance, that does not depend on our conscient will. Most memories concerning our experiences are hidden within our unconscious and it is not always easy to raise them voluntarily to conscience. All of us are subject to such a gap in the memory, which is mainly determined by the intrinsic limitations of our brain. We certainly know that our unconscious can make up, at least partially, for such lacks and, while we are fishing inside all the hidden experiential data in order to raise some of them to consciousness, in order to help solve any present problems, our unconscious may also—if considered necessary—make some arbitrary adjustments. So, even when we succeed in recalling some hidden memories, it may happen that they are presented to us with some variations, necessary to avoid dangerous psychological conflicts with our present social connections. In other words, while making up for the lack of our knowledge, our remembrances may be opportunely manipulated by our unconscious before inserting them within our Model Mentals Mental Models of Reality,[1] sometimes even at the expense of installing beliefs not necessarily supported by the reality of the lived events. Probably we will never get free from such false beliefs, since they will always try to stabilize themselves within our unconscious, thus allowing us to live with the certainty to have always been coherent with the role covered in our life.

As it is stressed by modern psychology, it is just our psyche that carries out such a pernicious mechanism (of which we might always remain unaware) only at the purpose of defending our mental integrity, by hiding or even changing, the meaning of some knowledge data in the case they could find themselves in a dangerous contradiction with respect to our present life, and therefore impossible to voluntary accept. Thus, it happens that we adhere to certain 'truths' or refuse them—sometimes even in cases in which they are respectively refused or accepted by the majority—mainly in function of our peace of mind, as requested by the stability of our role in society. Adhesion or rejection that, even when they have been plus or minus partially piloted by our unconscious, appear to us at all voluntary and justified, and to which we remain obsessively bound, so to defend it at any cost against other opposite 'truths,' as well as not demonstrable as ours, but that others defend with our same live and thickheaded belief, stabilized by the same unconscious processes, although in different social and psychological conditions.

०/०°०्०

The topics discussed in this book are of such a general type that invite the reader to maintain himself as far as possible from the way he finds himself when he is fully involved in the events of everyday life. Myself, more than tangling in

---

[1]Mental Models of Reality are discussed in the author's previous book, *Myth, Chaos, and Certainty. Notes on Cosmos, Life, and Knowledge*, Jenny Stanford Publishing, 2021.

the impossible knowledge of too many single details of life, I have always preferred to look at the world from the high in order to better observe the vastity of its connections without being personally affected by them and, at the same time, being fascinated by the normally unknown. All that, although being conscious of the extremely low percentage of the data that can be present to our daily attention with respect to that hidden in the immensity of the universe, and conscious also of the cost implied by the evident loose of many experiential data that could be useful during life.

To look from the top of a mountain, in fact, is not the same as looking from the edge of a street in a town. By looking from the high we lose a lot of details of the landscape in front of us but, fortunately, such a kind of observation, free from any physical and psychological implications, allows us to look and compare vast areas without being emotionally involved. To observe things from far should always be the correct way to adopt when trying to discuss any events without being personally involved, even considering that such a condition may contain the opposite risk to put in a second order some hidden important aspects about the facts under discussion. A condition, this one, that necessarily implies the need to resize our knowledge curiosity for anything falling within the human dimensions, but that stimulates the independence of thought, thus promoting deeper and less interested considerations on whatever we observe. A concept that has been authoritatively expressed in the literature both by Marco Tullio Cicerone in his *Somnium Scipionis* and by Johannes Kepler in his *Somnium.*

The above considerations may be understood only by fully and sincerely accepting to be infinitesimal grains of sand with respect to the grandiosity of the events happening within the immense universe, unknown and perhaps never completely knowable. An attitude that may help to better understand both the smallness of man and that of the restrict ambient where we live on Earth with the everyday rhythms and with the fixed, sometimes false, certainties of the human society.

A way of thinking, this one, that has always been somehow natural for me, an astrophysicist who has carried out a large part of his professional life by investigating the cosmic space in order to study those incredible events occurring within it, in particular the fact that the same formation of stars is nothing else than one of the many self-organization processes of matter, due to the same physical forces that work both at much higher and at much lower dimensional scales with respect to us. Processes that seem to me a clear indication that the appearance of life on Earth could have been just the result of the same evolutive processes of matter in forms always more complex, until reaching self-consciousness. This implies the idea that on Earth—as well as on any other plus or minus remote planet in the universe—it may continuously emerge and extinguish new and unexpected forms of conscient life, compatible with the different physical characteristics of each single star and its complex system of rotating planets, each with its gravity connected with its dimensions and its possible atmospheres.

Although limited, our experience on Earth—if opportunely purged from any influences of human proudness or of religious kind—tells us that matter can self-organize and produce life everywhere, just by using any available, although tiny, material resources in whatever physical conditions, both down to the enormous pressure of the oceanic deepness, as well as within the extreme aridity of deserts. Such evidence suggests to not blindly exclude the existence of life on any other planet of the vast universe, nor excluding *a priori* the presence of a suitable atmosphere. If our universe would be full of life, as I believe, with millions, or perhaps billions of other planets capable of hosting life, then, even if man on our Earth would not find the way to stabilize and survive to any possibly destructive events along our Sun's life, it is well possible that many other planets in the vast universe, by us presently unknown, could accomplish (or have already accomplished) the task of the life's continuity.

If we will succeed in coming out of the prejudice, at all unjustified, that the only form of intelligent life is the one existing on our Earth, it becomes natural and legitim to think about the possibility of the existence in many other far worlds (nowadays, in the past or in the future) of vital structures, even slightly different from ours for organization strategies of matter and for quality and quantity of involved chemical elements. Such a perspective could help us to stop any cold and static view concerning the drastic and negative perspectives of the Second Principle of Thermodynamics with its view of a universe without any scope together with the end of any kind of life.

ᒃᢙᵒᢙᵓ

Let me heartly thank my wife, Riri, and my son, Mauro, for their propulsive thrust, and Prof. Evgeni Borisowitsch Starikov for his preface and epilogue, which have made this work more precious.

I wish to thank also Mrs Jenny Rompas, Director and Publisher, Jenny Stanford Publishing, for her patience in waiting for the completion of this work.

Chapter 1

Just One ... but Dual!

Scientists are always in concern
if real fits with logic and extent.
With empathy an artist is in move,
transported by feeling and insight.
Everything, his being passing through,
the ascetic lives, embraces, and exploits.

About logics be careful my dear,
confidently all the savants say,
remember duality of brain.

—RB

From the Top of the Mountain
Rosolino Buccheri
Copyright © 2025 Jenny Stanford Publishing Pte. Ltd.
ISBN 978-981-5129-22-9 (Paperback), 978-1-003-50829-8 (eBook)
www.jennystanford.com

1.1 A Summary of Old, Classical, and Modern Views

Human history shows a continuous progressing of achievements of any kind. A progress that allows us to gain an always greater knowledge of the world around. Unfortunately, our smallness in comparison with the immense Cosmos does not make us aware of its enormous extent, of its complexity and, especially, of its evolutionary direction. Notwithstanding it, stimulated by knowledge needs, we tenaciously believe to be able to overcome any limitations.

Today, in our everyday thinking, we assume that what we perceive corresponds to the existence of real objects, independently from whoever is perceiving them and how its characteristics are seen by others. Many years ago, it was not so.

The Miletus school (7th–6th century B.C.)—considered as the start of Greek philosophy—was marked by the aspiration to discover the *phisis* (φυσυζ), the essential nature of all things, from all possible viewpoints: scientific, philosophic, and religious; disciplines that, at those times, were not so separated as they are now. For mistics there was no difference between spirit and matter, everything was animated, and all forms of existence were manifestations of the φυσυζ. Thales too saw the divine essence in everything—like the Chinese Taoism where nature itself manages the changing— and Anaximander conceived the universe as an organism furnished of a cosmic breath, something like old Saxons' view.

Heraclitus of Ephesus, instead, conceived the world as a perennial movement, a continuous transformation originated by the reciprocal action of antonyms' pairs. Parmenides of Elea, on the contrary, considered the change as illusory, and conceived the idea of an immutable substance, the *Being*, unique and undestroyable, that fills everything and causes the illusion of changes that we perceive in all objects. Finally, some cosmogonies refer to four elements (water, air, earth, and fire) as the origin of everything perceivable.

The concept of *atom* as the last undividable unit of matter from which all the things are made up (a further development of the human thought, introduced by Democritus and Leucippus) occurred very long before the advent of the scientific rationality, when the idea of spirit was separated by that of matter, and the movements of the last were attributed to external forces of spiritual origin. A dualism spirit-matter, body-soul, therefore, occurred well before the advent of the scientific rationality, that in the following centuries became an essential element of the future Western thought.

In the course of time, the progress of thought generated many important differentiations in all Western cultures. In the 16th century, spirit became an important field of application of religious and philosophic theories, mainly referring to tradition and moral, while the description of matter, made free from the Aristotelian doctrines' influence, became a specific task of science, due mainly to Copernic, Galilei, and Newton. At the same time, René Descartes expressed the dualism spirit-matter (*res cogitans* and *res extensa*), basis of the modern Western culture. With the Cartesian dualism, man was identified with his mind and the "I" became a pure spirit trapped in the physical body considered part of inert matter. A separation between mind and body—therefore between physical sensations and sentiments—bringing in us conflicts of all kinds (social, cultural, ethnic, etc.). Descartes expressed such concepts in a very radical form by affirming "I think, therefore I exist," meaning that mine is the only undoubtable existence when I observe and think, the rest (even also my past and future existence) is nothing then just only a construction of my mind.

Against Descartes' thought, the idealists Gottfried Wilhelm Leibniz and George Berkeley postulated that only what is thinkable is conceivable, and all what is unthinkable does not exist. According to them, the world existence, together with anything in, it is bound to their existence in the God's mind, from which everything we think and see depends, a concept not universally accepted and particularly contested by Bertrand Russell, as we may read in his *Authority and the Individual*.[2]

In the classical view of the world, typical of modern society, we assume the existence of an objective reality independent from its perceivers and endowed with intrinsic properties, a reality that we denote and explain through concepts, symbols and numeric values defined within a set of explicative theories. In other words, every physical system under observation may be described by a certain number of dynamic variables associated with well-defined numeric values that define the system at any instant. Starting from the knowledge of such a physical state and of all forces and constraints aging on it, classic physics was thought to be able to show the procedures apt to deduce the equations of its previous, present, and future motion. Such a way to look at the things, led Pierre Simon de Laplace to think that we should look at any state of a physical system as the effect of his previous state and as cause of the following one. According to this view, he wrote in his *manifest of determinism*:

> An intelligence which, for one given instant would know all the forces by which nature is animated, and the respective situation of the entities which compose it, if besides it were sufficiently vast to submit all these data to mathematical analysis, it would encompass in the same formula the movements of the largest bodies in the universe and those of the lightest atom; for it, nothing would be uncertain and the future, as the past, would be present to its eyes.[3]

[2]Cfr. Bertrand Russell, *Autorità e individuo. I doveri dello Stato e i diritti del cittadino*, Longanesi & C, 1970, Milano

[3]Pierre Simon de Laplace, *Essai Philosophique sur les probabilités*, Bachelier, 1840, Paris.

It must be noted that in Laplace's deterministic viewpoint is implicit a sort of 'theory of everything,' a view in which man is able to foresee all phenomena, known and not known, present and past, together with all their connections. This is equal to say that the universe might be visible, according to Laplace, as a unique block (*block-universe*) from its most remote past to its remote future. Views, these, that derive from the *exophysical approach*[4] of classic physics of the last millennium when it was thought that man may be able to reach a complete detachment from its ambient, so being able to observe and analyze it from outside without any disturbing mutual interaction.

The classic view, in fact, was based on the *scientific realism*, an *exophysical* line of thought which depicted a world made, from one side, by object and relations together with their intrinsic properties, and from the other by observers supposed to be able to identify such properties by means of experiments, without being affect by the world around.[5] Such an attitude of superiority with respect to nature has certainly allowed an enormous facilitation in the process of drafting the logic-mathematical rules able to describe and keep under control the natural phenomena, and has also allowed the formulation of the simple and elegant theories of classic physics, without which all the fundamental science and technique achievements, together with their significant spin-offs in every sector of our civil life, would have not been possible.

Classic Physics' achievements consolidated along time the trust in science, with the consequent idea that man had already reached a finally objective representation of reality. Such a trust was explicitly expressed in April 1900 by William Thomson (Lord Kelvin), who, by referring to the still open

[4]The concepts of *exophysics* and *endophysics* will be discussed in detail in Chapter 3 of this book.
[5]Cfr. Evandro Agazzi, *Varieties of Scientific Realism. Objectivity and Truth in Science,* Springer, 2017.

problems of the *ultraviolet catastrophe* and of the light velocity, told that "only two small clouds were still obscuring the clear sky of science." Lord Kelvin, convinced that these 'two small clouds' would be clarified very soon, refused to receive new students to train as researchers, by considering not in view any new research advancement. As we know today, Lord Kelvin was wrong: the 'two small clouds' were soon clarified just at the beginning of the new century by Max Planck with his blackbody theory and by Albert Einstein with the discovery of the light's velocity, a great conceptual revolution that radically changed the future course of science.

1.2 Quantum, Prigogine, and Contemporary Science

Soon after Lord Kelvin's simplistic overestimation of the situation — by the way shared by many scientists of that time — important hurdles were soon found in the scientific research. The first one was at the beginning of 1900, when Max Planck enormously amplified the first 'little cloud' by introducing the concept of 'quantum of action' that gave the start to quantum physics with the connected problems of counter-intuitiveness concerning the phenomena occurring at very low dimensional scale. Important examples are the *entanglement* that change from deterministic to probabilistic the cause-effect principle, the so-called 'problem of measurement,' the Bohr Complementary Principle, the Heisenberg Indetermination Principle, etc., topics that may be found in any modern book of physics.

The second hurdle happened soon after the result (incredible for that time) obtained by Michelson and Morley when measuring the velocity of light: a result that confirmed the Relativity Theories—both the special and the general ones—

published by Albert Einstein, thus unsettling all previous knowledge about light and space-time. Results that were followed by a conceptual upheaval from which a flourishment of new and fundamental research started, still now active. From such problematics new studies and elaborations started on the theory of systems, on dissipative structures, on chaotic and eco-systems, and so on. A flourishment harbinger of always new discoveries, among which we cannot forget the results obtained by Ilya Prigogine about the self-organization of physical systems far from the thermodynamic equilibrium. Very important research lines, these, revealing that the stability and the survival possibility of every living system—including man—depend on their virtuous exchange of information and energy with the ambient.

Concerning the theory of systems, let us recall here the 'emergence' of new phenomena impossible to foresee starting from the knowledge of their single components. Studies, these, that had triggered a very intense debate, bringing from one side to a constructive confrontation between the different schools of thought about the problem of the intelligibility of nature, but also to the feeling of somebody[6] that science may find itself in a crisis for which new paradigms need to be elaborated.

Within the contrast between the development of new research activities and the still-alive logics of classic physics, I find here interesting to cite, from one side Einstein's *realism* expressed in the famous 1935 article[7] where the authors, by considering the correspondence between theory and reality, assumed that reality may be known by means of the experiment, and from the other, the *interactionism*, to which so many thinkers like Niels Bohr, Ilya Prigogine, Karl Popper, Edgar Morin (although if somehow from different visual points)

---

[6]Cfr. Thomas Kuhn, *Dogma contro critica*, 2000.
[7]Cfr. Einstein, Podolsky, and Rosen, *Can quantum mechanical description of physical reality be considered complete?* 1935.

base themselves, i.e., the assumption that the unavoidable continuous interaction between man and the ambient in which he lives, if from one side assures to him a stable although temporaneous existence, it affects in a radical way his construction of theories and models concerning the representation of the world reality.

With the advent of quantum theories and those of non-equilibrium thermodynamics, the classic views concerning many old problems shifted radically, showing that linear processes are just an idealization unable to solve complex problems. Linearity, in fact, does not exist in nature where all real phenomena have always to do with situations of complexity at all levels and where chaos and unpredictability are at all normal. They emerge to our conscience in very complex physical situations together with the need to elaborate large quantities of data.

Chaos and unpredictability are not due solely to the inability of our brain to elaborate enormous quantities of data, they tell us about the condition of variability that we encounter due to our interaction with the ambient. A condition that pushes us to quote Gödel's theorems of completeness[8]—an important result of the mathematical research—concerning the overcoming of the claimed self-referentiality of the logic-mathematic laws.

Gödel's theorems show that the fundamental principles of any formal system must be taken, always and obligatorily, outside of the logic system itself. This admonishes us on the illusion about the existence of 'complete' formal systems, able to express final truths, regardless of any external interferences. At this purpose, Odifreddi says that

> no one theory pretending to found mathematics may ever be self-justifying, but is forced to look at a justification out of itself.[9]

---

[8]Cfr. Ernest Nagel and James R. Newman, *Gödel's Proof*, 1958.

[9]Cfr. Piergiorgio Odifreddi, *La matematica del novecento*, Giulio Einaudi, Torino, 2000, p. 47.

In other words, every reasoning, even scientifically very rigorous, may never be self-referential because it exists always, along any logic chains, a concept based on something external to it: something not necessarily demonstrable, possibly coming from beliefs or intuitions not always adequately expressible by means of the mathematic language.

Such a theorem excludes the existence of obvious truths out of those experimentally verifiable, and invites to be careful to consider as real 'truths' many of those that refer to common thinking, thus allowing the derivation of a series of consequences also considered true. Probably, in such cases the simple concept of 'objectivity,' is not clearly distinguished from the less simple one of 'intersubjectivity.' A concept that, being referred to persons that live inside the same social ambient and are then subject to the same experiences, may become relevant just for that social community, even if it does not contain any absolute truth from which to infer ontological foundations of the world.

I find it proper to underline here that Gödel's theorems translate into a scientifically rigorous form a consideration already expressed in the past, for example by Blaise Pascal in the 17th century, when he lucidly observed that all demonstrations base themselves on a series of concepts, logically interconnected but based on some primary concepts assumed implicitly, without any possibility of a demonstration within the same proposition. If, in these cases, in order to pretend to be 'objective,' we use the term 'demonstration' in its logic-mathematic meaning, typic of the scientific method, we risk falling into contradiction because no demonstration of objective truth may ever be reached outside of the level of empiric verifiability, safe anyway only under specific environmental conditions. Any 'demonstrations' obtained within empirically unverifiable ambits may always be polluted by subjective aspects, and therefore cannot avoid depending very strictly on the culture, on the experiences, and on the beliefs of those who follow the demonstrative reasoning.

Such a lack of autonomy of formal systems could be the cause of contradictions in many cases, particularly when we investigate microcosm and macrocosm, for which direct and final conclusions are not always possible, thus revealing the intrinsic limits of the scientific method if it is applied in fields far from our sensorial experience.

1.3 Uniqueness of Self-Conscience

The differences of perception and evaluation observed in different individuals imply that everyone of us has his own, specific, way to relate to his ambient. Therefore, the concept of uniqueness of self-conscience and of its development along time must be referred to each single person, in which it evolves along a well-defined direction, certainly different from that of all other individuals. It is a matter of fact that the awareness of each individual way to be and to think is the result of the development of any single person in consequence of his struggle for satisfying his needs, even against all possible contrasting conditions. This implies that any single self-conscience is necessarily bound to the specific way of working of the connected perceptual system, able to analyze the stimuli coming from all sensorial organs distributed along his body, with the chief task to satisfy his own vital needs and, at the same time, adapt them to the external ambient.[10]

As the self-conscience, also the concept of independent external reality forms itself starting from the analysis of the stimuli transmitted by the sensorial organs along our life.

---

[10]I find interesting to remind that Fritjof Capra was convinced that the self-conscience could be the result of the self-organization process of the intrinsic characteristics of living systems along the adaptation process with its ambient. Cfr. Fritjof Capra, *The web of life* (*La rete della vita*, 1997).

The methodology of such an analysis consists in constructing different classes of stimuli based on the observed differences along the everyday life. In general, the better sensitivity of a sensorial organ with respect to another one may (unconsciously) imply the attribution of a major value to stimuli coming from the last. This, because our sensorial organs have not the same importance for all of us, nor within ourselves. We know, for example, that we can 'feel' the external ambient even in absence of sight and hearing, although within the intrinsic limitations of the perceivable optical and acoustic frequencies. There is no doubt that every single person evaluates in his own way the comparison between the pleasantness and the painfulness of a perception in order to attribute importance to it. In other words, every difference in the selection of stimuli admitted to the cerebral analysis may produce corresponding evaluation differences among different individuals in their way to represent the external reality. We may easily agree on what above as soon as we look at the differences between diverse populations or—even if less marked—between different individuals of the same social group.

Concerning the analysis made by our brain of any signals arriving from sensorial organs, we may assume that the general methodology may consist in constructing mental classes of objects and events based on the observed differences in the perception of any single stimulus. The result of such a process is the awareness of the existence of forms, colors, dimensions, and distances that, by the way, are perceived in a slightly different way by any single person, even with 'normal' sensorial organs. It is evident from what above, that, even acknowledging the general characteristics of what we observe, it is also true that any even small differences, both in the selection of the external stimuli and in the interpretative logics used, may produce evaluation differences, plus or minus important, among single individuals—and even more important, among different social groups with different traditions—in their way to represent the external reality. Differences that have always been observed, and can always be verified in the everyday life of different peoples, as well as within the same social group.

1.4 Rationality, Intuition, and Common Sense

Today, our knowledge about the structure of the universe is at a level never reached before. The most of directly observable phenomena, including many of those concerning the atomic and subatomic world, or those connected with the immensity of the Cosmos, are explained with great details from the logic-linear structures of mathematics and physics. Relativity Theory is able to predict many important cosmic phenomena like quasars, neutron stars, *Black Holes*, and gravitational waves, just to quote the most known of them. Quantum Physics (QP) is nowadays on the basis of all the modern technology, especially electronic and chemical, together with the coding, transmission, and decoding, of all those data needed today in our civil life. Extraordinary successes, these, both in terms of knowledge and for the consequent technological uses, that have highly strengthened the credibility of science and the subtended belief on rationality. Today, due to the achievements obtained along the last century, the concept of rationality is identified with the scientific method on which the logic and mathematical techniques are based, together with their experimental verifications.

From the beginning of the last century, the investigation about the phenomena occurring in the macrocosm (the immensely large that we see but cannot handle) and in the microcosm (the immensely small that we do not see but that we may only control their observable consequences at our dimensions) have caused an upheaval of our common sense.

In particular, the deep investigation of the infinitely small has posed our science in front of important contradictions in terms of both thought and language. When studying the phenomena occurring inside the atom, and sometimes also those occurring in the immensity of the skies, the usual visualization method coupled with our common sense does not help us anymore as it does for what happens at our dimensions, and let us risk to wrongly interpret our observations and measures, even if we often consider them precise and unequivocal.

The achievements of QP already discussed, as with other important statements, like Gödel's theorem, or any other discoveries in the fields of neurophysiology and of complex systems theory, are few concrete examples where we clearly see the limits of the classic vision: they tell us that we cannot always separate the observing subject from the observed object. As a matter of fact, just referring to experiments done at the atomic level, we see very clearly that matter does not find itself in well-defined places—as classical physics claims—but it shows only a propensity to be in a certain place with a certain probability, then posing the problem of the precise localization of objects. The Heisenberg Indetermination Principle tells us in addition, that by increasing the precision with which we know the position of a particle, the measure of its velocity becomes always more indetermined, this implying a limit to the simultaneous knowledge of all the properties of a particle. Finally, the very fact that electrons or light may manifest themselves as waves indefinitely dispersed along space or as perfectly localized in a specific point in function of the method with which we observe them, stimulated two great scientists Werner Heisenberg and Niels Bohr to say that what we observe is not nature as it is but nature exposed to our investigation methods. According to quantum physics, in fact, the concept of physically isolated particle is an abstraction, since their properties can only be defined and observed by means of their interaction with other systems. Quantum

Physics, in fact, has practically dismantled the classic concept of object as independent and locally defined, and has indicated that we cannot talk of nature without mentioning who is observing it. So, QP reveals to us the fundamental unit of our universe and shows us that we cannot decompose it into minimal units endowed with independent existence and that all material objects falling under our eyes are not distinct entities as our common sense would suggest but that they are inseparably bound among them, with the ambient and with us, their observers.

By considering that nature is a net of relations among everything existing (including human beings), we derive that we never see the ultimate essence of the reality but only its interactions with the rest of the world and with ourselves, part of it. The very fact that at our macroscopic level we clearly distinguish any single object from any other, is simply because the probability to precisely localize an object increases with its dimensions, and since at our dimensions we are in contact with objects whose dimensions are like ours, their localization is practically certain. In a few cases, however, physics tells us about unity even at a macroscopic level, in contrast to our common sense. We know, for example, that in Einstein's Relativity Theory, space cannot be separated from time, matter cannot be separated from its gravity field, and gravity field cannot be separated from curved space. According to Relativity, then, matter, space and time may be seen as parts of something inseparable.

1.5 Duality and the Creative Tension of Antinomies

Plato described the antinomic condition in which man finds himself, by saying that

Man is double because of an intrinsic dualism, both ontological and gnoseological, since the corruptible knowledge
of the earthly dimension collide with the uncorruptible
knowledge of the Hyperuranium.

He was referring to the antinomic condition that, after
many centuries, Blaise Pascal would define a "monstrosity of
contradictions that can never satisfy." Today, an explanation
of such an antinomic condition is given from neurophysiologic
research. They tell us, in fact, that duality is one of the most
fundamental anatomic-functional properties of . the human
being, particularly, of his brain. The two cerebral hemispheres,
even being substantially identical from the anatomic viewpoint,
and even being constantly connected to the cortical and
subcortical levels, manifest evident functional asymmetries.

Elkhonon Goldberg says that with the emergence of
rationality, the human brain has chosen to work with two
cerebral hemispheres, characterized by the fact that the right
one is more active in the man's childhood (both filo- and onto-
genetic) when models and schemas are absent, while the left
one is more active in the adulthood when he systematically
uses schemas and models already acquired. From the physical
viewpoint, this equals to mention the specific cerebral
dominance recently intervened (in an evolutive sense) and
localized within the Broca and Wernicke language areas,
respectively in the left frontal and temporal lobes of the
brain. The action of such a dominance is that of having shifted
our knowledge from *Mythos* to *Logos*, at the start of man's
rationality. Such a complex situation is indicated by Antonio
Damasio with the existence in us of a 'nuclear conscience,'
connected to the philo- and onto- genetic start of human life and
an after intervened 'extended conscience.'

The nuclear conscience refers to a kind of knowledge able
to look at any external situation in a global way, somehow
unknowingly, probably bound to primordial instincts, but

anyway able to allow the storage within our unconscious of a great number of experiential data that we can retrieve rapidly and intuitively, without any rational action. It is, as Damasio says, an activity done unconsciously, but that sometimes allows the solution of vital questions, otherwise unresolvable due to the enormous time needed for their conscious analysis. Probably, many known ingenious intuitions made by artists and scientists, or even some sublime spiritual manifestations made by mystics and saints, may be connected to the 'nuclear conscience,' generically characterized as 'irrational.'

The other side of the coin, however, is that such perceptions, even able to contain important truths, may not be scientifically demonstrable by a linear logic, with the consequence that they are considered 'irrational' and subject to personal interpretation. Anyway, beyond any possible cultural schema to which these perceptions may be connected, this is not the same as to say that such expressions of the mind must always be considered as due to paranoia, therefore seen as 'false beliefs,' according to the definitions given by James Hillmann.[11]

Along history, still according to Damasio, our nuclear conscience has continuously grown, tending toward a system capable of a rational elaboration of all single and instantaneous units of nuclear conscience, a system from him called 'extended conscience,' a term that should include a selection of all experiences already done, schematically organized, and localized within the cerebral cortex, and communicable by means of our linear apophantic language.[12]

The 'extended conscience,' by using the logic methodologies and categories constantly refined along time, expresses the logic modality of knowledge, by us called 'rational' and today identified with the typical methodologies and categories of our

---

[11]Concerning "false beliefs," Cfr. James Hillmann, *On paranoia*, Eranos Jahrbuch, LIV, 1985.

[12]Antonio R. Damasio, *Descarte's Error. Emotion, Reason, and the Human Brain,* (*L'errore di Cartesio. Emozione, ragione e cervello umano*, Adelphi, 2008).

modern science. All this—together with all the possible positive and negative connotations and cautions about its correct interpretation—may show that at a certain point of the human history a double possibility was opened to our ancestors, i.e., together with a *mythos*, made up of intuition and empathetic disposition, a new alternative arose, consisting in a rationality made up of logic, calculus and schemas, even without cut ties with the same *mythos*, that has continued to work in order to reach an always more complete comprehension of reality.

As we have already shown by citing Pascal and Gödel, we may recall here for completeness that our formal knowledge of reality is still now mainly based on a logic chain of prepositions always interconnected to premises or basic principia not always and necessarily of rational type, and therefore, that also the modality, generally called 'rational,' may often contain negative characterizations which are bound to preconceptions not necessarily demonstrable, or to primordial instincts not appropriate to our social living, like egoistic individualisms finalized only to interests of single individuals or of entire populations at the detriment of others.

This is because very often the 'rational' modality depends 'rationally' on assumptions often derived by the irrational modality, with the consequence that it ends with maintaining till the conclusions of the 'logic' reasoning, its negative characterizations there contained, generally without being aware of it. It is a matter of fact that even if the starting of a reasoning is polluted by initial assumptions not necessarily true but so assumed in the current way of thinking, very often its conclusions are also assumed as true if considered rationally valid in its development. This pushes us to maintain constantly our attention about what we think is true.

Let me conclude this topic by establishing the inescapable fact that the two knowledge modalities—the irrational one typical of art and the rational one typical of science, the first addressed to the present and to the shadowy participation with the ambient and the second addressed instead to an organized control of it—coexist in all of us in a dichotomous, sometimes lacerating tension, opposite and irreducible to each other but also full of concrete possibilities. In order to positively use both modalities, however, it is necessary to accept and overcome the unavoidable tension that accompanies their simultaneous presence. Any lack of equilibrium inherent to such a tension may cause negative psychic phenomena, even the removal of one of the two opposite tendencies, like a unilateral development of the consciousness bound to primordial instincts, or oscillating in a cyclothymic way, or, finally, coolly calculating. The brilliant examples given by the great personalities who have guided the positive path of our history illuminate us on the way to face such a problem. Looking at their vicissitudes, we may observe that when man has been able to grasp and foster the *charge differential* of such a tension, taking aliment from the derived conflictual unease, the energy from it generated has always created new horizons of knowledge.

Chapter 2

Second Principle of Thermodynamics: Is Any Door Really Closed?

From the Great Explosion all was in:
billions of galaxies with life,
then, disperse matter going away,
swiftly running to nameless lands,
where worlds, lives, or expertise
to unusual emptiness will die.

It's just what knowledge today
suggest for the events in their flow,
between the known and what we still don't know.

—RB

From the Top of the Mountain
Rosolino Buccheri
Copyright © 2025 Jenny Stanford Publishing Pte. Ltd.
ISBN 978-981-5129-22-9 (Paperback), 978-1-003-50829-8 (eBook)
www.jennystanford.com

2.1 From Simple Hydrogen to Complex Life

Modern cosmology may explain with the abundance of details the evolutive history of the universe from its start until today, but about the very questions of its origin and its end, we do not have yet precise answers. In the following, I shall try a summary of what we claim about the events that gave rise to the origin of the universe and of its walk toward the complexity, then to the formation of our solar system and to the appearing of life on Earth by means of the continuous transfer of energy from the Sun, after the formation of our planetary system.

Cosmologic studies suggest that our universe could have been born 13.7 billion years ago from a singularity i.e., from an infinitesimal lump of matter of extremely high density and temperature, a physical state indescribable using the today known physical laws. Starting from this original 'singularity'—emphatically called *Big Bang* by Fred Hoyle—our Cosmos would have expanded and, just after a few minutes, there would have appeared all the basic ingredients by which it is made of, and from which all the structures we know today subsequently emerged: nebulae, galaxies, and galaxy' clusters, together with other extraordinary phenomena like pulsars, quasars, Black and White Dwarfs etc., including the living beings at least on Earth. The starting singularity is described by our theories as a single 'object,' a clump of a highly symmetric 'cosmic proto-plasm' containing *in nuce* all the properties of the matter that would subsequently emerge with the *Big Bang*. The theory suggests that its symmetry would have been broken just following the explosion, thus bringing to the emersion of Time and the subsequent expansion of Space, together with many other structural specificities today known, as it has been previously mentioned. Along the further evolution, some of these specificities became always more complex, while others disappeared in order to give life to always new structures, while their interposed space continuously increased.

Let us give a brief look at such evolutive processes, a more detailed exposition of which can be found in Guido Tonelli's book *Genesis*, when he writes about the 'great novel of the origin.' The basic ingredients of the just-born universe were initially neutrons and protons that, in mutual equilibrium under the action of the fundamental forces (mainly weak and strong nuclear forces), structured themselves in hydrogen (H, three parts) and helium ($He_3 + He_4$, one part), together with some atoms of deuterium (H_2) and very little Lithium (Li). Soon after, under the action of gravity, a very large quantity of such elements formed the stars from which, through the nuclear fusion processes, always more complex atoms progressively emerged, and among them, also carbon, oxygen and nitrogen, very essential for the following development of vegetal, animal, and human life. Both the fundamental forces and the elementary particles emerged from the *Big Bang*, are characterized by specific numerical constants (the electron mass, Planck's constant, the atomic energetic levels, the gravitational constant, etc.). According to current theories, the values of such constants have been fixed once and for all at the birth of the universe, thus conditioning its subsequent evolution, the one that we know today rather than others that could have been bound to different physical-chemical conditions unable to originate life. It is like to say that the physical situation emerged from the *Big Bang* is 'tuned' with life as it has evolved from the simplest forms up to humans able to study their origins.

In the enormous amount of time that has passed from the *Big Bang* to nowadays, the contrasting game between the gravity and the initial expansive impulse has given rise to an intense hive of organizational activities of matter, from which billions of galaxies were born, each one containing hundreds of billions of stars of any type and size. Today we observe a greatly dynamic scenario where mutually attracting galaxies destroy each other and restructure themselves by giving rise to new celestial configurations, while gigantic *Black Holes* slightly engulf them, thus continuously renewing the whole space. In the meantime, within any single galaxy, little and large stars are continuously born and transformed, and die through colossal explosions, slightly or rapidly according to their mass. The ashes of the dying stars dispersed by such transformations accumulate in the space and continuously enrich themselves with always new materials that substitute the exhausted one, so giving rise to new generations of stars, in a continuous cycle that, very slowly compared to human times, modify the general scenario of our universe toward a still unknown conclusion. In such a general change context, also our Sun, star of average mass in the immense cosmos, originated by some of the already cited dramatic events, will slowly modify itself together with its group of planets, comets, and meteorites, ending as such in order to generate new structures. Today, our Sun is at one-half of its life—in the so-called 'main sequence' of the Herzsprung–Russell diagram—and is still now transforming into helium its internal hydrogen mass, which was accumulated in the past due to gravity, or due to the thrust derived from some violent explosion in the near space. A process, this, that is common to all stars in the universe: their life, although flooding the Cosmos with light, is anyway limited, at least due to the quantity of the burning oxygen available. A limitation that is also referred to the life of the whole system of planets around it, whose necessary energy is ensured by the star.

Concerning our solar system, in 4–5 billion years from now, when all its nuclear fuel will have run out, our Sun will become an enormous *red giant* star, capable of embracing all its planets, including our Earth, and everything else it will have produced until that moment, life included. After that, the Sun will contract within itself thus becoming a *White Dwarf*, and then it will slowly switch off, while much of its external matter, scattered along the nearby space and subject to the same gravitational attraction processes, will go to feed other processes of stellar and planetary formation, as above sketched.

Not all the stars, anyway, have the same evolutionary history. For stars greater than at least 1.4 times our Sun, our theories suggest the destiny to quickly collapse within themselves, thus becoming very dense and steeply rotating objects, called *neutron stars*, some of which are seen as *pulsars*, since they emit pulsating energy toward Earth's direction. We have observed pulsars in binary systems,[13] ultrafast pulsars catching energy from the company star thus increasing their rotational speed,[14] or even pulsars rotating around another pulsar.[15]

Very large stars, much larger than our Sun, are instead destined to collapse into enormous *Black Holes* that could slowly absorb and swallow by gravity everything around them; certainly, sparse matter, but also stars, planets or even entire galaxies. It seems that this process has already started within our galaxy, where a giant *Black Hole* at its center is today slightly swallowing all nearby star systems. Let us remember, also, that the life of stars, even supporting human life for millennia by flooding the Cosmos with light and energy, is limited by the quantity of oxygen burning inside them. Consequently, the life of planets turning around them is also limited.

Concerning our Earth, at the end of a certainly very long period—but like a flash with respect to the expansion time of the universe—in one of the innumerable processes of matter's self-organization, life emerged. At the beginning, we had only unicellular organisms, ancestors of all present bacteria, able to feed and reproduce themselves. Such organisms have evolved by means of a gradual process of self-organization,

---

[13]Russel A. Hulse, Joseph H. Taylor, 1974.
[14]Valentin Boriakof, Rosolino Buccheri, Franco Fauci, 1983.
[15]Marta Burgay et al. 2008.

going toward always more articulated and diversified forms, until the appearance of mammals, followed by primates and finally by man and his culture. A complex organization of live and dead matter that, in principle, should also be extinguished in less than a couple of billion years, when all the Sun's hydrogen will be completely burned together with its emitted light and heat, needed to support life.

Life, anyway, could disappear even before, if in the meantime some huge, always possible, catastrophe caused by planetary collisions could intervene, as it happened sixty-five million years ago. At that time, an enormous boulder wandering in the space impacted our Earth, thus provoking a stirring on the mainland, in the seas, and in the atmosphere, so as to inhibit the vital activities of all living beings of relevant physical dimensions such as dinosaurs (that were, in fact, totally extinct), but also to luckily save the life of many smaller beings (like us), so favoring the development of the human civilization development. Nobody might exclude new possible reasons of disasters in the future, today unpredictable, that may cause important self-harm to life on Earth.

Stability and transformations on Earth and in the Cosmos

By looking at the man's history along centuries, it is not difficult to convince ourselves that the physical structure of human beings has remained approximately the same for millennia, while our single lives have gone on, one after the other, in an acceptably stable world. Each single person perceives any changing situations in his/her life by means of the stimuli transmitted from his sensorial organs and of the consequent analysis made continuously by his brain. Stimuli that every one of us perceives in his own way in consequence of the specific needs to which he/she is subject along his life, and to which he responds with the actions considered adequate for any specific circumstances. In other words, every single person organizes his/her living process in such a way to manage at best the interaction with the external word, being plus or minus positively stimulated by the variable circumstances of life, as for example important physical or relational events, especially concerning family, ideologies, religions, or traditions.

By considering all kinds of differences between us—as due to physical structure, to culture, to social conditions, etc.—we might infer that the external reality, perceived, analyzed, and acknowledged as independent from each single perceiver, is seen differently from each of us, therefore remaining unknown in many details, with the consequence that nobody of us may imagine a reality exactly equal to that of any other person.

Such differences of evaluation are very much amplified as soon as we turn our attention to what happens in the enormous universe, particularly if we are plus or minus conscious of the ceaseless and colossal events occurring there, and if our senses, being accustomed by millennia to calibrate our representation of reality based on our daily experiences, can judge them in their intrinsic complexity.

Humans' life on Earth is extremely short in comparison with the durations of the astonishing changes taking place in the far space. As a result of such an enormous difference, the continuous occurrence of the huge stellar changes and catastrophes in the skies are not directly observable by us, with the consequence that the heavens appear to us immutable with all the constellations fixed in the sky and considered not able of any relevant effects on the continuity of the human's life on Earth. A similar reasoning holds for all those sequences of events telling the history of our planet through the very slow transformations occurring on the terrestrial surface, like the very long modification rhythms of the terrestrial geography, known by studying fossils and geological events. If all of us could have the possibility to fully understand the difference between the enormous extent of the processes occurring in the expanding universe and those occurring on Earth's tiny scenery in which we live, we would factually realize the meaning to be on an extremely small and far place sited in an extremely peripheric position of one among myriads of mean size galaxies of the immense universe, a place that appears to us very central, as it would be in any other position within any other celestial structures. It is a situation, ours, for which the duration of human life, of the order of one century, appears to us very long, while it is just a flapping of a butterfly's wings if we compare it to a star's duration from its birth to its death. The flapping of a butterfly's wings, anyway, within which everyone exists, lives his own emotions every day, accumulates knowledge, observes his own ambient and thinks about his origins and those of the world.

Within such a general stability of the physical and climatic conditions of Earth, and based on the idea that the world in which we live is the only available for still many thousands of years, man has realized and programmed activities of any kind wherever on Earth, and promoted exchanges of information, both among single persons and through official Institutions at this purpose established, in order to be able to know in any possible details whatever is needed to live at our best.

2.2 Evolution, Emergency, Mind

If we agree to include in the term 'evolution' the expansion of the universe—both concerning its totality and all its single structures continuously modifying themselves along time and space—we may imply with this term also the formation and the structuring of all the single components of the universe itself, including the formation and the following advancements of the human and animal life on Earth and, implicitly, also any kind of life that might have appeared and evolved on other planets of the universe illuminated by a star, even if we would never know it due to their enormous distances from us, impossible to investigate today with any possible tool. We may further convene to include in the term 'evolution' also the progression of the complexity with time of any structure, including that of the humans' physical aspect and of the positive changes of the human knowledge with the consequent advancement of culture and the ability of man to control events. Properties, the last, that emerged from the matter's self-organization along the human brain formation. An evolution that has allowed us to become observers and analyzers of the world.

All this, in the consideration that, coherently with the expansion of the universe and with the appearance and development of life, our knowledge, too, evolves, together with our ideas about the way to manage all the situations occurring within our society in a continuously changing world. An evolution that is generally an advancement, for which what before may have appeared as wrapped by the aura of the myth or of the hope might become a real datum or dissolve into a dream and disappear in order to take new and more favorable ways of knowledge's growth.

This is just as it happens today, when we consider how the world, subdivided into many communities by centuries, is now exploding into a system totally interconnected. A world for which we could easily foresee the future disappearance of many different local traditions with their implicit pseudo-certainties that, protected until now by the difficulties of communications, are destined to converge—certainly after an amount of time very long with respect to a single human life—toward a more common and ampler worldview and, therefore, toward a new common *modus vivendi* as the effect of the universal communication relations.

2.3 Time's Flow and the Increasing Brain Size

In what we have discussed until now, it has always been implicit a concept given for granted—the flow of time—a concept that did not exist before the *Big Bang* but that since then it has become a very essential element for the description of any phenomenon, certainly concerning the development of the universe but also and especially for man's life on Earth. It is then of particular importance to briefly review our idea of Time and of the way we feel its flow, concept that we all should consider of basic value for anything we may further discuss.

The sequence of natural phenomena, like night and day, the seasons' changes, the cyclic presence of celestial events, as well as other events pertinent to humans, like coming to life, social living, working, and dying, are all events that induce man to think about the flow of time. A flow from which, according to our subjective perception, we may be differently affected, in function of our age, of our physical state, or even in function of the rhythms of our daily activities within family and within society. Effects that we might attribute to the specific way of working of the man's central nervous system with respect to our variable activities, inside our family and in the social life, by considering that any pleasure or grief we may feel is always bound to the biologic rhythms characterizing us within our small dimensions, rhythms so much different from those characterizing the universe's long walk from its evolutive start.

Human life lasts on an average eighty-ninety years and within such a tiny time interval we learn how to understand and control our world. Other living beings (animals or plants) bring to completion their lives in slightly less or slightly more time, but anyway always enormously small with respect to that of our universe. Nevertheless, even within such extremely small durations, we can witness a continuous transformation of any phenomenon, living or not living, without any hope to return to the previous state.

The concept of 'time' is very controversial in some respects and very easy to accept for others. It is true and acceptable, for example, that along the average human life, those of us who do not suffer from particularly psychologic problems and are not subject to any strong drogues, conceive time as a unidimensional reality of which we can only control the present, not certainly the past because it is already established and it may partially be remembered. About the future, we may just hazard hypotheses based on how we live the present and based on what we remember of the past; sometimes we may even affect the future according to our needs and hopes, but we have not the power to completely control it.

The consciousness about the flow of time, anyway, is not equal for anyone of us and it could even show important differences between individuals, sometimes even within the same person in different physical or psychical conditions. Time's flow in some cases may appear as slowing down or, on the contrary, to flow too rapidly, depending on our age or in special psycho-physical conditions. Moreover, besides these very common situations, the medical literature shows us very many other less 'normal' cases, in which particularly anomalous descriptions are given about the perception of the time flow. Cases, these, perhaps due to the conditions of the perceiver (psychologic problems, assumption of drugs, practice of the profound meditation, etc.), as often described in the literature. In these cases, the flow of time may be described in strange ways, either very much expanded or contracted, or discontinuous, or even at all fragmented. In more extreme cases, it can even be perceived as completely stopped or, on the contrary, in continuous expansion. A great variety of perceptions, then, as the result of our personal contact with the external world, a contact that cannot be exactly equal for anyone of us and that could be explained with the different abilities of our brain to connect the various events of life.

Unfortunately, the same physics, even being the most sophisticated scientific discipline concerning the description of the 'objective reality,' is not able to explain the reason for the time flow from past to future and, even more so, it cannot explain all these perception differences.

A notable attempt to describe in detail such phenomena has been done by Susie Wrobel's studies.[16] I find it interesting to cite Metod Saniga's effort to describe the temporal aspects of altered states of consciousness by means of a simple geometrical schema, linking their mathematical and physical features with psychology, psychiatry, and philosophy.[17] A construction, the last, that uses the language of the projective geometry to describe in a symbolic way, both the so-called 'normal' perceptions of the time flow together with the most extreme groups of pathologic perceptions, like the sensation of absence of time, the stop of the time flow or the immersion in the past.

Some researchers today, by trying to justify the enormous difficulties to insert the arrow of time in the equation of motion in physics, are convinced that time does not exist. Albert Einstein himself wrote that "the flow of time is only a stubborn human illusion." We may ask ourselves if the present ambiguous way we feel and describe the flow of time might be due to today's limitations of our brain, unable to keep in mind the myriads of information continuously acquired along life.

---

[16]Cfr. Susy Wrobel, *Fractal Time. Why a Watched Kettle Never Boils*, 2011.

[17]Metod Saniga, *Algebraic geometry: a tool for resolving the enigma of time?* 2000.

It could also be that the ongoing evolutionary process of the formation of our sensorial and cerebral behavior (related to the principle of cause and effect that allows the serial ordering of events) is nowadays at a stage not yet concluded. If this would be true, the flow of time characterizing our life today could only be the result of the still imperfect evolutionary status of living systems, unable to represent the lived reality in all its aspects. If this would be the case, based on the present knowledge about the tendency of our brain to increase in size,[18] we could expect in the future a growth of the human mental abilities, able to slowly reduce the importance of the passage of time. In this case, the volume's increase of the human cerebral cortex—place deputed to the superior activities of the thought—would then evolve such to include always more fragments of memory in the brain, so giving the possibility to describe with always greater precision the past and, at the same time, increasing the vision of the future, much more than it has always been. Such a situation would approach the already cited Laplace's dream, at least concerning the past and anyway remodulated by quantum indeterminations, and it would be shown that Parmenides was not at all far from the truth when he said that nothing changes and that everything, both past and future, exists always, at each moment of our life.

If this would be the case unless the disintegration process of our planetary system would be faster than the increase of the cerebral capacities of the human brain, the concept of time as we know it today could consistently change. A change that could even be accelerated in the next future—even beyond the evolutionary process of the human

---

[18]It is important to recall here the discovery of the gene NOTCH2 that, 3,5 million years ago originated three new genes able to triple the number of our neurons. Let us notice that besides our Earth, an ongoing brain increase could also interest the living beings of other planets in the universe.

organism—by the technologic progress that gives us always more sophisticated means able to refine our perception and analysis of the external ambient. A process that, in agreement with the continuous increase of information that we continuously get from the outside, could be able to improve the view of the future, thus giving a great help to the always more detailed and urgent actions to program in order to reach our always more complex objectives.

It is interesting to notice at this purpose, that Ilya Prigogine, while studying the systems far from the thermodynamic equilibrium, understood the need to review the concept of time, left unexplained by the classic physics because of the difficulty to describe its characteristics of directionality. A review needed in order to build a 'historic' and therefore 'human' science, like all other disciplines.

The important research activities in biology about evolutionism, and Prigogine's work on the dissipative structures, together with the connected concept of irreversibility, have strengthened the concept of directional time, thus making even clearer the classic physics' failure in this respect. Science is a human product and so, according to Prigogine, it cannot have a *status* different from all other disciplines, then pretending to show more 'truth,' even if we cannot exclude its technical superiority in developing operational instruments able to describe and control reality.[19]

In any case, the highlighted differences of perception about the time's flow in all 'normal' persons, together with the vast available knowledge about the altered states of consciousness, are not certainly compatible with a uniform Time without any privileged direction, as described by the classic physics, independently of the assignment (from outside) of a precise arrow of time.

---

[19]Cfr. Ilya Prigogine, *The end of certitudes*, 1997.

2.4   Intelligent Life in the Expanding Universe

As a premise to this chapter, let me strongly disagree with the opinion of many[20] that life may exist only on our Earth, an opinion that I find devoid of any acceptable scientific justification. We have still many holes of knowledge concerning the universe, particularly about its birth, about its future development and about the end of its expansion. Conscious of such ignorance, I will try some hypothesis about the development of life (in the future and everywhere in the enormous universe), considering the existence of myriads stellar systems of all kinds, very many of them certainly not so different from our solar system, then able to guest life. Stellar systems, these, that perhaps we will never have the possibility to know in detail, thus putting the claim about the absence or presence of life there within the group of the religious or atheistic beliefs. The universe is so large with respect to our small but haughty will to judge, that all assumptions about the existence or non-existence of life in such an uninvestigable place are at all arbitrary with respect to any consideration free from dogmatic assumptions. If we succeed to escape from the prejudice that the only possible life may be on Earth, it becomes lawful to think of the existence in many worlds—certainly very far from us and from themselves—of vital structures, today, in the past or in the future, possibly slightly different from us concerning their physical forms and dimensions, or concerning quality, quantity, and organization of involved chemical elements. Our fantastic literature in the scientific-informative sector is full of such a possibility not always excluded by scientific argumentations. Based on these considerations, in the following, I shall talk of 'intelligent

---

[20]Cfr. Ex. James Lovelock, *GAIA. A New Look at Life on Earth*, 1987, and Marcelo Gleiser, *Il neo del Creatore*, Rizzoli, 2011.

life' instead of 'human life' wherever it may occur within the immense universe.

We may imagine that in case of a very long trip in the space, looking for refuge on some far planet, the physical structure of travelers will have to change in order to adapt to the new existence conditions in the limited ambient of a spaceship for very long times, in absence of gravity. In such an adapting process, especially concerning the very high speed expected to be reached by the future spaceships during the search for new living places, Time shall have a relevant responsibility due to the changed conditions to which our body will have to adapt, especially because of the slowing down of our biologic processes dictated by the Relativity Theory. Today, accustomed to living on Earth, we do not give the correct importance to such topics because our experience, genetically transmitted by millennia from previous generations, is implicitly binding us to a well-defined paradigm of the world representation. A paradigm that is characterized by the implicit thought that even what we do not observe and do not know must necessarily possess forms and behaviors equal to those known today on Earth.

With all this in mind, let us discuss any possible consequences on the future of intelligent life, starting from the hypotheses—consistent with the present cosmologic views—for which the total quantity of matter today present in the universe could not be able to stop, or even invert, the gravitational escape impressed by the *Big Bang*. A condition, this, which allows the continuous expansion of the universe, together with the mutual estrangement of galaxies and the consequent progressive annulment of their gravitational attraction. It is believed that in such a condition, life would cease for all those vital structures that should find themselves at distances so large from their stars to forbid any further evolution. The entropy in such a case could continue to increase but only until reaching a value for which matter could not aggregate anymore, a final value corresponding to the 'thermic death.'

Let me object to some aspects of such a belief. First, it would need a very large number of millennia for a decisive removal of galaxies from each other, such that the mutual gravitational attraction would not be felt anymore. More than that, if it seems reasonable a mutual estrangement among galaxies in the foreseen escape condition, it is less clear the mutual estrangements between nearby stellar systems within each galaxy, and even less clear a mutual estrangement within each single planetary system, as a result of its removal toward the unknown. In practice, the death of old stars and the birth of new ones could continue for eons, thus saving life on Earth and on any other planet within any stellar system for an enormously long time.

In practice, concerning life on Earth or on any other habitable planet, this would mean at least the next four billion years allowed by the presence of oxygen. So, the question is: isn't it much too early to start worrying about the danger of incommunicability among single galaxies, and especially within single stellar systems of the same galaxy? Concerning our Earth, if we just look at the increasing human cognitive skills, it does not appear at all uneasy to think that the intelligent life could undergo important modifications, significantly in time to avoid the worst possibilities!

We do not know, of course, until which point and at which velocity our brain might grow, but in the light of the evolution of the species enunciated by so many great biologists—beyond any interesting disputes among Lamarckians, Buffonians, etc.—it would be a nonsense to immediately close every door, at least theoretically, on what the future may reserve to human life. It would not be a great idea to deny *a priori* a possible and decisive evolution including a further increase of the human brain before that the distances between the celestial systems of the universe would possibly reach the no-return point![21]

Personally, due to the little knowledge available today on such topics, I consider it interesting, other than scientifically and humanly lawful, to dream the fast and continuous improvements of the cognitive abilities of intelligent beings on any habitable planet of the universe. A sort of optimistic arrow of time in virtue of which 'humans' could be able to keep in mind an always higher number of vital experiences before arriving at a gravitational situation of no-return. A 'humanity' with enormous inventive resources, therefore, able to easily wander through immense spaces using for example quantum means on which it is today difficult to make precise technical previsions but that do not appear at all prohibitive considering the status of present knowledge.

Concerning the case where the matter of the universe would be insufficient to continue the get-away from gravity and started to invert its motion in order to go back to a *Big Crunch*,[22] it is almost impossible to try any kind of previsions about the future of humanity. In such a case, both the concept of entropy and those of life and future would miss their present meanings. To these questions, nobody of us may have any answer based on our today's knowledge, and the present human pride to preview everything may only be delivered to a dream or to religion.

2.5   Entropy: A Law or Just a Principle?

This section discusses the conceptual error derived by having allowed to let rise the Second Principle of Thermodynamics up

---

[21]This prevision is based on the unlimited expansion of the universe presently foreseen, but not yet certain, as we will see in the following!

[22]A circumstance not believed today by many, but it cannot be at all excluded *a priori* because of the insufficient knowledge today available on the amount of total energy present in the Cosmos (see Section 2.8, *Thermal Death or 'Intelligent Design'?*).

to a Law, therefore with the characteristics of full universality—an error that has got us to the point of allowing too extreme assumptions, somehow of religious type, on the future of the universe and of human life.

At the beginning of the 19th century, for the great importance gained by vapor machines in their development along the Industrial Revolution, a study was undertaken of the temporal behavior of all physical systems, starting from the simplest ones, those made up only of gas molecules. Since Thermodynamics was the scientific discipline dealing with such a topic, its Second Principle was considered, conceptually and technically, of relevance. Following the procedures of the gas' kinetic theory, the behavior of gas was analyzed by means of a theoretical experiment consisting in the mixing of two sets of equal gas molecules, disorderly moving within two boxes at different temperatures separated by a partition wall and considered isolated from the rest of the world. Let us stress here the importance of the fact that the molecules of the two gases are not individually visible by the humans performing the experiment, and therefore we may only see the difference between their initial and final states.

If the partition wall between the two boxes is thought to be eliminated, the molecules of both gases will reciprocally collide with each other and finally will mix up. Because of the repeated collisions among the molecules of both gases, the theory correctly says that the faster molecules of the warmer set will soon communicate part of their kinetic energy to the slower ones of the cooler compartment and then, after a few minutes, since their motion is at all casual, the excess of energy of one of the two parts will equally distribute among the molecules of the entire container, thus reaching a common degree of thermal agitation. As a result of such a mixing, the gas will be distributed over the whole recipient with a temperature averaged with respect to the mean temperatures of the two initially separated compartments, and then the previous order, consisting in two distinct sets of gas at different temperatures, will be annulled. The theoretical result following from such a procedure is that from now on, there will be no more heat to transfer from the first recipient to the other, the disorder's level will have reached the maximum possible value and the molecules' chaotic motion will remain forever statistically the same.[23]

The theory concludes that in such a process, as in any other process of gases mixing in a recipient isolated from anything else, the gas molecules of the process cannot return back spontaneously in order to reconstitute the original system, because the energy used to destroy the original order has already been transformed into degraded energy—heat—consisting in the uniformly disordered motion of all the single molecules of the gas.[24] From this point on, any new increase of complexity should be considered as occasional and destined to soon disappear toward a final state that cannot be inverted anymore.

The above-described two-gas-system transformation may be seen as the passage from an initial situation partially ordered—because marked by the original separation of the gas into two separated parts at different temperatures—toward a final complete disorder which characterizes the indistinguishability of the two gases within only one recipient where the gas will not spontaneously separate anymore into two different sets. A similar conclusion may be thought when mixing more than two gases or liquids, or when mixing any number and kind of diverse gaseous or liquid substances. Milk and coffee, for example, mixed up in a cappuccino, will never separate spontaneously, more so if we would melt there a third substance, i.e., sugar. In all these cases, the theory and the

---

[23]Let us notice that the implications of the human intervention needed to eliminate the dividing wall between the two compartments is not discussed in the theory. It will be done later here with other considerations.

[24]The full procedure is described in any elementary text of physics.

practice say that the energy used for the mixing has degraded into heat and therefore it is not anymore available to retrieve, by any means, the originally separated elements. Such an increase is quantitatively measured by using the *Entropy*, a physical quantity able to measure the disorder of any isolated physical systems.

The above evaluations, derived by the simple theoretical experiment shown above, have been extended to any real process of mixing, on Earth as well as in the entire universe, as we will see in the following. In all cases, the resulting disordered motion of molecules is considered as continuously increasing in time, and any possible new increase of complexity is considered occasional and destined to soon disappear toward a final situation consisting in the final complete irremediable disorder that cannot be inverted anymore. A condition, the last, in which a stable and ordered aggregation of matter, even one able to favor the evolution of life — the maximum expression of order and matter aggregation —is not allowed anymore.

The increase of molecules' disorder derived from the results of the simple theoretical experiment described above, has been considered valid for any other similar mixing process of gas molecules, even if occurring in the entire universe, therefore even in any process of stars' formation, with the consequence that the entropy of the Cosmos is seen as destined to increase with time until bringing it to the thermodynamic equilibrium—like that of the two-gas mixing described above—and therefore to the end of any form of order, life included. A conclusion that is found to agree with the expectation that the present universe's expansion should bring to the transformation of all present stars into lumps of compact matter very far from each other, therefore not anymore gravitationally interacting.

It needs to be stressed that the process of gravitational amassing of hydrogen and helium molecules inside stars cannot be considered equivalent to the storing of gas molecules in a recipient on Earth, as proposed by Al-Khalili.[25] It is certainly true that the entropy of the hydrogen and helium molecules within stars is quite stable for a lot of time (even if at a value of entropy much higher than within a recipient on Earth), but such an enormous gas accumulation due to the very high pressure generated within stars is radically different with respect to what happens within a recipient on Earth. The enormous pressure due to gravity within stars is able (as explained above) to promote the formation of a great number of new chemical elements of higher atomic number, effect impossible at the low atmospheric pressure on Earth. Some of these elements (oxygen among others), dispersed along the sidereal space, aliment nearby planets provoking, within their atmospheres, the life's birth of plants and of animals of any kind, humans included.

To have forgotten the enormous difference of density and pressure between Earth and a star, the evolution of life has not been taken anymore in consideration and the temporal path of the entire universe has been seen as going toward a situation in which any further increase of complexity will have necessarily be limited to the proliferation of *Black Holes*, perhaps together with many other individual or complex units that, disorderly dispersed in the space, will become separated by enormous distances due to the estrangement of galaxies, no more gravitationally interacting, thus establishing the final fate of the Cosmos. A process that seems apt to bring us toward a final state where any new thermodynamic transformation, useful for a further complexification of matter, will result impossible to realize, so indicating the end of animal and human life.

If these were the things, it seems that the absolute disorder would finally win, so bringing the Cosmos to the announced 'thermic dead,' a situation that can be described as a cemetery of lumps of energy—more or less elementary—spread

---

[25]Cfr. Jim Al-Khalili, *Il mondo secondo la fisica*, Boringhieri, 2020.

within an enormous empty space. Based on such considerations, we are seen as going toward a no-return point when the full universe will be just a shroud of *Black Holes* extremely far from each other in which all galaxies should respectively concentrate and where life could not develop anymore.

Last century, the faith in such an inexorable increase of entropy pushed Sir Arthur Eddington, one of the most important physicists of last century, to affirm with punctilious proudness:

> I believe that the second principle of thermodynamic is at the highest place among all other laws of nature. If somebody would let you notice that the theory of the universe, from you considered the most correct, does not agree with the equations of Newton and Maxwell, worse for these equations. If we discover that such a theory is contradicted by experimental observations, you might think that the experimenters often do mistakes. But if it comes out that your theory is against the second principle, there is no hope for you: the only think it may happen is that your theory must miserably collapse.[26]

Eddington was an important theoretic physicist, certainly at the same level as Albert Einstein. On the other side, same as Einstein, he did not consider that the scientific research—with the connected world view—can never be considered completed, even with his very important works. Moreover, he was convinced that just the "pure thought could succeed in obtaining a complete description of the physical world"![27]

Eddington could not acknowledge that if it is true that the Principles of Thermodynamics are the result of observations done on Earth, and that they have never been contradicted by

---

[26]Arthur Stanley Eddington, *The Nature of the Physical World* (*La natura del mondo fisico*), Laterza, Bari, 1987.

[27]John Barrow, *The Constants of Nature. From Alpha to Omega* (*I numeri dell'universo*, Oscar Mondadori, pp. 84–85).

the working of the thermal machines, it is also true that there does not exist any experimental demonstration, nor any mathematical theory asserting their universal validity without any possibility of error. Something that, instead, Philip Morse acknowledges when writes that

> The laws of thermodynamics have a negative quality that let us distinguish them from all the other physics' laws, which makes very difficult any possible experimental proof,[28]

and, when he highlights the negativity of the Second Principle by saying that

> the lack of any direct proof is still more noticeable than for the first law

declaring, therefore, that it is impossible to demonstrate it on a theoretical basis. We may notice such a difficulty only based on our experience under terrestrial conditions: an experience that seems to have never given us a contradicting example, and therefore confirmed by traditions and beliefs without any imprint of universality.

2.6   Is the Analysis Complete?

Let us come back to our experiment done on Earth. The random motion of an enormous number of equal gas molecules, too small to be individually visible—both during and after the process of mixing of the two separate samples of them at different temperatures—is a situation that humans cannot control but just observe through their statistical behavior. The single particles are too many and too small to be singularly distinguishable by man, especially along their chaotic movement,

---

[28]Philip Morse, *Thermal Physics*, 1964.

thus giving us the possibility to observe only their equilibrium state at the beginning and at the end of the mixing process. Then, considering their continuous chaotic movement within the recipient, for us to hope to see the spontaneous appearance of the ordered condition existing before the mixing, we do not have other possibility than waiting until such an ordering may appear by chance. Unfortunately, as already said, due to their enormous number and their indistinguishability, we could never notice it. It could happen accidentally very soon or it would take an enormous amount of time, a duration that could be even greater than the entire universe's life—well beyond any human possibilities! So, we can only say that the initial ordered configuration may be, in principle, statistically reproducible but we cannot take it into consideration from the practical point of view.

A similar condition may be easily recognized in all other mixing problems, for example in the case of the milk–coffee mixing, where the time requested to pour coffee into milk and then mix is little and measurable, while it is not possible even to have an idea of the enormous time requested to separate the single ingredients of the mixture, individually invisible to us.

Such an ambiguous result of the statistical analysis leaves us uneasy with respect to the apparent net result obtained by the kinetic theory. We find ourselves in front of a situation which leaves us without a definite answer about the statistical approach, thus inducing some doubts about the described full procedure and, consequently, about the final fate of the entropy assured by the Second Principle. There is a suspect that something may be wrong somewhere in our analysis, either before the process of the gas mixing, or in the hidden, statistical, by man uncontrollable, dynamics of the random collisions along the transition from order to disorder.

Let us try to get some answers to such topics by reviewing in detail all the steps of the procedure, in order to know whether something has been underestimated. We observe that, in practice, some kind of molecules' accumulation somewhere within the recipient is always possible in a mixing process, effect that can be due to the unavoidable presence of physical forces, in particular, gravity and even the although weak nuclear forces, without talking of the human intervention along the various phases of the experiment, always and everywhere present but never considered until now. In view of such lacks, and trying to let the statistical viewpoint be coherent with the rest of the analysis, let us look at the full process again, including all physical forces, both long-range and short-range, always present, and not always necessarily negligible, that have been *a priori* excluded.

Let us start by looking whether it is possible to neglect— as the kinetic theory does—all the work done by man when preparing the experiment before mixing the two samples of gas at different temperatures. As a matter of fact, they must have been obtained in advance by means of some preliminary preparatory work made by a human operator, work until now unconsidered. A work that should have consisted in (a) preparing two separate nearby boxes, (b) obtaining two separate sets of molecules with different temperatures (starting, for example, from a single set) to put into any of the two prepared boxes, and (c) eliminating the separation between the two boxes in order that the two sets of molecules may mix. In these steps of the experiment (neglected in the theoretical experiment), some human energy has certainly been spent, but never considered, and certainly not easy to calculate. After the mixing of the two partitions, therefore, the official analysis has been considered as completed in view of the humans' impossibility to operate any kind of selection within a sample of too small, invisible, molecules.

Let us now think to include in our calculations the energy contribution of both the experiment operator and that of any other forces present, both along the process of preparing the experiment itself and of mixing the two samples of molecules at different temperatures. Then, let us think to compare this energy with that requested for the inverse process of individually separating the elements of the mixing, necessary to reconstitute the situation existing before the start of the experiment. Doing so, we become aware of the indecipherable difference of work requested by the human activity between that of the mixing process and that of the inverse process which aims to reconstitute the original situation.

Such a comparison forces us to recognize that we have included in our calculations only what we are able to quantify with our possibilities, and have forgotten what we are not able to quantify (due to the very limited possibilities of man to manage the extremely small), thus remaining unable to understand the limits of our evaluations. A forgetfulness that, together with that discussed above on the production of the heavy elements within the Sun, has led us concluding to agree with the Second Principle of Thermodynamics which thus has been raised up to a general physical Law by which we decide to make predictions on the future of human life!

Let us finally observe that, unlike the 'laws,' the 'principles' should only be considered as guides, hired based on their experimental evidence.[29] They should allow to make previsions about new phenomena (on Earth) or develop new descriptive paths about already observed phenomena on which a theory may possibly be based. At this purpose, I find it interesting to observe that Jeremy Rifkin, in his book *Entropy*, by describing the way toward the final disorder, writes:

---

[29]From the Treccani vocabulary of Italian language: *For experimental sciences a principle is a statement that constitute the generalization of a vast experimental evidence [...] being referred to the experimental evidence distinguishes it from the postulate (the simple premise of a hypothetic-deductive system) while the amplitude of its application field distinguishes it from a law.*

> In the universe everything starts with a value and a structure and irrevocably it moves on a situation of chaotic and dispersive randomness [...] when an apparent order is created in a whatever point of the universe or of the earth, it happens at the expenses of an even greater disorder caused to the surrounding ambient...,

and concludes by taking refuge in the religion to avoid any further discussion by saying that

> ... we must underline that the entropy law concerns only the physics' world where everything is finite and where the living beings must walk their path until when they cease to exist. It is a law that governs the horizontal reign of time and space. But it remains silent as soon we get to the vertical reign of the spiritual transcendence.[30]

Evidently, Rifkin, prefers to turn to an unprovable cause (God) in order to justify with an unverifiable claim, the ineluctability of the Second Principle. It is just the human presence with its powerful possibilities to handle very complex phenomena and, on the contrary, with its very limited possibilities to manage the extremely small, that renders us unable to understand the limitations of the Second Principle. Perhaps it is better to consider this Principle an important rule on Earth, as it was initially done, not a universally valid law on which to base even religious considerations!

2.7   Human Actions Produce Work, as Other Forces Do!

I found it interesting to reiterate, within the present discussion about the Second Principle, the fundamental notion that any mixing of solids, liquids or gases originally separated, is

---

[30]Jeremy Rifkin, *Entropy* (*Entropia*, Baldini and Castoldi, 2000).

done by the joint action of nature's forces and by man who, even if is helped by special conditions of freedom with respect to any kind of natural constraints, is part of nature too and, as such, he increases the complexity of the universe. Let us then look to any possible reasons of unease with respect to the categorical affirmations, strongly defended by Sir Eddington, Jeremy Rifkin, and others, about the spontaneous tendency of all physical systems to go ineluctably toward the thermodynamical equilibrium, then to the thermic death of all livings wherever in the universe, such to put the Second Principle within the group of most important nature's laws. This is like to say that when in chaotic motion, matter is not able to actively 'feel' the presence of any material forces that anyway exist always and everywhere, and therefore that heat, i.e., the disordered energy of the motion connected to the temperature, always prevails over the 'ordered' energy of the active forces—as it happens with passive forces like friction—so as to bring with time to their complete elimination. An assertion that should be applied to all kinds of forces, even to those forces coming from man's activity, because man, being the result of the same physical forces operating by millennia, is present and contributes to the processes of which we are just discussing.

At any rate, trying to understand if we have neglected some important detail, let us widen our gaze up to the enormous sidereal spaces in order to look at the universe in its totality. Doing so, we notice that it becomes impossible for us to consider absent or even negligible any of the natural forces (gravitational, nuclear, electromagnetic, ...) that have contributed to the formation of the universe and to its further complexification. A process that started with the presence of only atoms of hydrogen and helium and was followed by all other simple elements and compounds of any complexity until the living beings, that exist and contribute, same as all other forces, both to order and to disorder.

We have learned from our cosmological studies that nature's physical forces have favored the complexity's increase of matter until the emergence of human beings, a very complex force of nature who, being able to understand when the other natural forces are favorable or opposite to his force, may contribute to the further increase of order and complexity. We observe from forgetful time that the universe proceeds toward always more complex developed and elaborated energy and matter states, through the constant growth of structures. It is enough to think of the spontaneous transformation that, starting from the 'primordial broth' of particles born with the *Big Bang*, has led to the formation of always more complex forms like galaxies, stars, and planets, till humans. A continuously growing process, never interrupted for more than four billion years, that, starting from those primordial cells derived from the 'emergency' of the informative property in the inert matter (from which life derived), gave birth—at least concerning our small and perhaps insignificant planet among the billions till now unknown and perhaps lost within the immense universe—to an indiscriminate flourishment of living structures, plants and animals until human beings that populated and continue to populate our Earth.

The informative property alone could not be enough, anyway. By looking at the production process of carbon within stars, we easily notice that it is also possible to confute the assertion that thermic energy can never be used! As already discussed above, carbon's production within stars implies a process of resonance where the same thermal energy of the star becomes fundamental to allow the formation of large quantities of this important element, fundamental for the existence of human beings whose actions are not considered in the formulations of the Second Principle.[31] Such a neglect,

---

[31]Cfr. Section 2.1, *From Simple Hydrogen to Complex Life*. The topic is discussed in detail, for example, in R. Buccheri, *Myth, Chaos, and Certainty. Notes on Cosmos, Life, and Knowledge*. Jenny Stanford Publishing, 2021.

together with that of the presence of all other natural forces, particularly those of attractive type like gravity and short-range nuclear, constitutes the most important problem concerning the *status* of the Second Principle. Furthermore, in view of such evidence, nobody, without any prejudice—of religious nature or of any other type—might ever exclude that such a process could have happened (or may happen in the future) in other billions of sun-planets systems present in the universe, possibly with other possible forms of life with respect to those on Earth, in dependence of the physical characteristics of each planetary system.

To summarize, we may convene that our universe has developed from the *Big Bang* starting with only hydrogen and helium that, by means of gravity, accumulated in large quantities forming the different-sized stars of our universe. Thereafter, under the propulsive thrust within stars, all other more complex chemical elements arose. Elements that, expelled from stars' central nuclei, may be found in the planets, together with all the derived complex substances like solids, liquids, and gases, followed—certainly on Earth—by plants, animals, and human beings.

So, any compound of any complexity, human being included, arose from the evolution of a confused amalgam of simple original elements, generated by the same natural forces that originally arose soon after the *Big Bang*. Therefore, if we consider—as we should—that humans, as any other living systems, are the result of the constructive encounter between nature' forces and elements along the extremely long chain of complexity improvements, then the previous affirmation equals to say that the evolution of every aggregate of simple elements until human beings is the natural evolution of simply hydrogen and helium within stars.

For what we may understand, this continuous increase of complexity—from simple atoms to humans—that started after the universe's formation, seems to have been followed by a period of general dynamic stability in which we live today, with living structures who continuously self-reproduce, and dye. An incessant transformation from disorder to order and vice versa, therefore, considering that even the wastes of the vegetal, animal, and human life, other than disorder may also constitute the possibility of order, as, for example, food for any other kind of inferior living beings, or fertilizers, as previously cited. We should not consider at all improbable that the inclusion of man's operativity within the game of all other active forces, until now neglected, could radically change our view on the trend of entropy in the universe, in partial disagreement with the Second Principle.

All that above tells us that it is not possible to generalize *tout court* to the entire universe and to the life tenaciously developed within it, the process of the mixing of two gases of simple particles originally separated in two different recipients and not subject to forces! We cannot do this without considering the enormous difference existing between simple particles not subject to forces and complex systems like chemical compounds, or even living systems, able to use at their advantage a field of forces, not talking about the enormous forces acting within a star!

Perhaps, we could fall into believing it if we look only at the life of individuals on Earth that ends with the death and the decomposition of their own body. If, instead, we consider the life transmission from the father to children occurring since eons on Earth—and foreseen to continue for still many centuries, perhaps even in more complex forms with respect to the present ones—the things do not appear anymore so simple. We could discover that, once born, life does not end anymore; rather, it becomes always more conscious of the forces to which it is subjected and more resistant to them! In this case, we might consider the Second Principle not effective in the long run all over the universe, and valid only within the limits posed by the classic views, a wonderful construction that, on the other side, has clearly shown since time its limitations. How can we be so sure that life is only a temporary phenomenon, even so immensely long and always growing in complexity? We could only believe it if the present theories about the foreseen end of the universe could be considered established and final. But, as we have already specified, we have still so much to discover about what concerns the start and the possible end of the universe!

2.8   Thermal Death or 'Intelligent Design'?

The unsettling perspective of the entropy's constant increase and of the consequent thermal death, prospected by those who have looked at science for decades as the theater of the chance and probability, is nowadays opposed to the idea of *intelligent design,* prospecting the progress of nature toward the structuring of always more complex forms of conscious life addressed to the formation and diffusion of intelligence, as indicated by an always greater number of scientists. It is easy to observe a general tendency to deny the possible existence of such an optimistic arrow of time, opposite to the pessimistic one derived by the Second Principle and directed toward the final death of humans together with that of the entire universe. It is possible that such a tendency is derived from the fact that the agreement about the Second Principle has been already consolidated for more than a century in our minds, while our knowledge of the concept of increasing complexity is still at a rudimentary stage. However, if we consider the continuous growth of human intelligence as an objective and inevitable fact, we should review the pessimistic aspects of the Second Principle that until now have appeared as inevitable to many of us.

Obviously, the human cerebral growth, from which the increase of the intellectual abilities of humans derives, is independent of the transitoriness of the life of each one of us, whose body perishes and shrivels toward disordered matter after a certain amount of time, on average in less than a century. But, when we consider the transmission of life from parents to children, the situation changes, and we discover that life appears as continuously evolving toward increasing complexity and always more resistant to any forces that would defeat it. On Earth, through the coupling man–woman, such a transmission lasts for millennia, always increasing in complexity, a circumstance suggesting, at least in principle, that life could continue toward intelligent forms always more complex and able to control the future.

Somebody could object that life on Earth (as it could be on any other planet) is anyway limited because it depends on the light and energy furnished by its star. We surely know that the life of stars, even being able for millennia to flood the Cosmos with carbon and oxygen, is not eternal due to the limited quantity of hydrogen that can be burned. But it is also true that our Sun will continue to shine for another four billion years and, because of the energy continuously transmitted to Earth, nothing prevents that during such an enormous period man could succeed with his increased cerebral abilities to invent new and faster means to travel along space so to have the possibility to colonize other planets and to extend the continuity of life toward other sufficiently close stellar systems. We are not in conditions to exclude if such a condition is already going on in other remote stellar systems that, being out of our possibility of observation, will never be visible to us on Earth. The many unknowns today present in our knowledge should let us think that the future of life in the immense Cosmos is still an enormous rebus!

One of the most important of such unknowns concerns the amount of matter in our universe. Our present knowledge tells us that the ordinary matter in the Cosmos is only 5% of the total existing matter, while the remaining 95%—the so-called 'dark matter'—is made up of a kind of elementary particles, still of unknown nature, concentrated for 27% within galaxies, and distributed for the remaining 68% everywhere between galaxies. Besides the fact that it permeates all the space, the precise properties of this last kind of matter are not yet precisely known, but could be extremely important in order to have more precise indications of the evolutive future of the universe. We do not know, in particular, if the expansion velocity of the universe (that depends just on its total mass and therefore also on the 'dark matter') is such to finally interrupt the present run toward outside and eventually return back, as it happens with a tossed object. The expansion of the universe appears today unlimited, but the situation could change if we would discover that the mass of this still unknown dark matter is sufficiently large to interrupt the headlong rush impressed by the *Big Bang*, such to reduce its escape velocity until annulling it, or even provoking the return toward a *Big Crunch* of the whole matter into one enormous *Black Hole*. For what we know today, the only possible answer we may have for the future of the universe is the *Black Holes'* evaporation by means of the quantum tunneling effect proposed by Hawking,[32] a process that could have the effect of replenishing the cosmic space toward a, yet undetermined, physical state. Summing up, nobody maybe sure today about what will be the universe in its remote future, which suggests avoiding blindly basing ourselves on the current knowledge about the Second Principle, whose general validity has been (perhaps arbitrarily) extended to the entire universe.

In any case, even if we have still so many uncertainties about the universe's remote past and future, we know about the present constant growth of its structures which constantly proceed in time toward states of matter and energy always more complex and elaborated. It is enough to look to the spontaneous transformation that, starting from the primordial

---

[32]Cfr. Stephen Hawking, *The Theory of Everything*, 2003.

simple broth of energy and particles coming out from the *Big Bang*, has driven the formation of galaxies, stars, and planets, and then many forms of life until the most complex one, today represented by the human being and by his continuous extension, both as single units and as a complex society. Such a growth process that, starting from the emergence of the *informative property* of the simple matter, has populated our Earth of life has never been interrupted over four billion years, thus allowing the development of life in a continuous flourishing of living structures, plants, animals and finally humans. Nothing forbids that this may have happened also in other planetary systems, perhaps in different forms with respect to the way it has happened on Earth, according to the physical characteristics of each planet within the immense universe.

Such a regular path toward the appearance and the evolution of mankind establishes a continuous unidirectional advancement toward the *diminishing* of entropy. A path that might be named an 'optimistic arrow,' a condition that many scientists today tend to deny on the basis of the Second Principle's previsions, whose pessimistic horizons appear to many unavoidable possibly because, as sometimes has happened along history, many are convinced to have everything clear in mind, or because they are implicitly bound to a mystic view of the human life, as if life wouldn't be a natural phenomenon consistent with nature's laws, but as if it would depend on an *ad hoc* divine intervention, destined to finish toward nothing (for some) or toward an afterworld of souls to judge on the basis of their life on Earth.

If we want to avoid hiding ourselves behind a finger, we must consider not only one but two arrows of Time, one in a deadly competition with the other. One of them, inherent to inert matter, pointing in the direction of degeneration and entropic disorder, and the other which points toward an always more conscient life, fundamental as the first and with a not less subtle meaning, even if its final goal is wrapped in the mystery of our ignorance.

Certainly, for a hypothetical system of simple particles not reciprocally interacting and moving chaotically within a fixed space (especially a recipient) and for which the presence of any kind of force—both attractive and repulsive—is excluded, we may certainly agree with the expectations of the Second Principle. But, may we say the same thing in the case of complex systems, especially if their elements are subject to forces, both reciprocal and external and, more, if they go continuously to increase in complexity until becoming intelligent and able to use blind nature's forces? If we consider that life's complexity tends to grow with time together with its social and technologic creations, is it lawful to transfer to the entire Cosmos, and particularly to living systems, the statistical concepts of thermodynamic, valid for simple particles not reciprocally interacting and in a chaotic motion, isolated from the rest of the world? Even with the many important unknowns of our knowledge, nobody may be authorized to affirm with safety that the evolutive process that along some billion years has brought to the development of life, first in elementary forms and then, little by little, toward always more complex forms until humans, has necessarily to stop and that life should extinguish for a presumed continuous increase of the entropy!

Unfortunately, the enormous difference between the duration of cosmic events and that of the human civilization— even more, that of our single lives—prevents us from thinking that our physical and mental abilities (that we believe as absolute and immutable) have still an enormous time available to continue growing by means of new and perhaps unattended 'emergent' properties. Such a condition is unfavorable for letting us become aware of our present low level of evolution, thus preventing us from imagining how could be in the future the physical and mental capacities of human beings. Is it not so strange to hypothesize that, by having still available a few billion years before our Sun becomes a red giant and destroys Earth, human beings could have already acquired all the techniques necessary to travel along the sideral spaces, looking at the opportunity to find safer places where to live and continue to improve his cerebral and organizational abilities. If we succeed in widening our view toward a more remote future—perhaps very far for us who write and read these notes, but certainly compatible with the remaining, still very long, period of habitability of our Earth—nobody might exclude *a priori* that man's mental and physical abilities could evolve toward much ampler views and powers of execution, such to succeed to use at their most all his experiences and knowledge, today prudently hidden in our unconscious, thus finding new and more favorable ways for a deeper understanding of all what surround us. All that for an always more useful exploitation of our natural resources, other than for more evolved ways to travel safely along interplanetary spaces, in order to colonize other livable worlds, before that life on our Earth would become impossible due to the dying Sun, thus increasing the possibilities of further evolution of intelligent life.

Today, we consider it obvious and natural to embrace certainties of various kinds that we consider always and ever valid because we have observed their stabilization and fortification for time periods that to us appear enormously long. But, the consideration that the time period in which we are able to do concrete evaluations on the existence of our civilization is extremely short with respect to the cosmic times in which the full universe is developing, should invite us to caution when giving judgments of eternal validity to our, even firm, beliefs, and to understand that we are too small entities for being able to understand completely the enormous resources of a Cosmos whose vastity of spaces and power of change is so far from our daily worries! Conscious of the space-temporal limits in which we are constrained because of our physical and astronomic conditions, we should try as much as possible to come out of them at least mentally, thus avoiding being trapped in fixed beliefs, porter of intellectual stasis, and in any unmotivated persistence on preconceived ideas, opposite to the concept of evolution.

Such a waiver would allow us to look without any binding prejudices to the possible future of the humanity, especially concerning any possible new knowledge achievements, so giving to the quote *homo sapiens,* the required fullness of meaning!

2.9 Will Always Be Valid All What We Believe Today?

In case of a secular trip in the space looking for refuge in other worlds, the physical structure of human beings will have to radically change in order to adapt itself to the new existence conditions, either for very long times in the limited ambient of spaceships in absence of gravity, or when landing on other planets and needing to adapt to their specific atmospheres and gravity forces. In such an adapting process, concerning the extremely long travels toward the search of new living places, Time shall continue to have a relevant responsibility because of the changed conditions to which our body will have to adapt, for example, due to the slowing down of our biologic processes dictated by the Relativity Theory, but not only. Today we do not give the correct importance to such topics because our experience on Earth, genetically transmitted by millennia from previous generations, is binding us to a well-defined paradigm of the world representation. A paradigm that is characterized by the thought that even what we do not observe and do not know should necessarily have forms and behaviors at all equal to those known today on Earth.

The clearest example of such a tendency is the attribution to God of moral thoughts, and sometimes even physical characteristics, which are equal to ours, even recognizing in Him a Being that transcends the human nature. This, independently from the fact that religiosity, seen as the hope that something of us will remain after our death, is a feature that unites almost all human beings, to whatever religion they may be bound. Given for certain that we will never know anything about what may happen after our death, it may be usefully comforting to refer to the Entity that we call God, even if His characteristics end always with being associated to the culture, traditions, and related mental filters of the social group to which we belong.

Even atheists, anyway, ask themselves the same questions, to which they cannot or do not want to give answers, either because they think not being able, or because they do not like to blindly entrust to unknown external figures the world management, as for example does Lee Smolin, for whom

> The idea that dominates everyone is that the rationality responsible of the coherence of all around us is not in our world but is hidden beyond it,

thus, entrusting the evolution of the universe only to nature laws, whatever the word *Nature* may mean.[33]

On the other side, as Michał Heller says:

> One thing are the mathematical equations describing the nature's laws, another thing is the nature that follows them: who, or what may instill life to mathematical equations?[34]

---

[33]Cfr. Lee Smolin, *The Life of the Cosmos* (*La vita del Cosmo*, 1998, Einaudi, Torino.

[34]Michał Heller, *Creative Tension. Essays on Science and Religion*, 2003, Templeton Foundation Press.

From whatever side we consider the question, we can only see ourselves as products of nature, and as such, we might never know what nature exactly is and Who or What governs it. As part of nature, anyway, we represent the world at its highest level, on the consideration that we are its most complex products, able to think about ourselves at the point to even discover the laws from which we derive.

It may be interesting to remember here that some physical theories propose that the universe could be itself a quantum automaton able to manage itself,[35] a superorganism, of which we all are parts, living beings and inert matter at different levels of complexity, all contributing to the flow of events which defines its evolution along time. In any case, even if the universe would be itself a superorganism, it could have its own objectives for its evolution and then, all what today exists could be functional to its existence. If this would be the case, man himself would be the most complex and powerful instrument to reach these objectives, and then the sentence 'nature wins always' would not be a simple fatalism but the consequence of the fact that man, as the most powerful and sophisticated instrument of the same nature, might be able to bring nature toward its objectives, maybe through positive and negative fluctuations, but always along a generally positive progress.

---

[35]Cfr., for example, Buccheri, Jaroszkiewicz, and Saniga, Endophysics, the fabric of time and the self-evolving universe, 2003.

Chapter 3

Integrative Considerations

To the unknown, fleeing, I wander;
to every landing, new fast escapes,
and lingers on the shores of the why.
Plenty of questions to each answer,
and climbing peaks for better see;
but higher the top, lesser discern.

On the knowledge's enchantment we talk,
but youth's certainties and hopes
may agonizingly dye and leave us in pain!

—RB

From the Top of the Mountain
Rosolino Buccheri
Copyright © 2025 Jenny Stanford Publishing Pte. Ltd.
ISBN 978-981-5129-22-9 (Paperback), 978-1-003-50829-8 (eBook)
www.jennystanford.com

3.1 Intelligibility and Hidden Aspects of Reality

Elkhonon Goldberg, in agreement with Damasio's research, reminds us that when our mental schemas are not yet well-shaped,[36] the first kind of knowledge is of global-empathetic type, i.e., based on the global and immediate examination of the data, not yet rationally analyzed, and not yet interpreted, therefore only referred to a subjective view. Only after infancy such knowledge is rationally analyzed, interpreted, and communicated by means of the language. At any rate, because of the above-discussed physical differences that characterize us, not all the knowledge we acquire along our life may be communicated in exactly the way in which it has been acquired. This is just to repeat that the communication can never have the character of objectivity in its strict sense. In other words, the social consensus may never eliminate all possible elements of subjectivity both in the communicants and in the recipients. Therefore, the necessity to bring the communication to the personal dimension of every one of us for its pacific use in the context of the common sense, may result in an unconscious concealment or modification (surely different from person to person) of crucial aspects of the communication itself. Despite it, different social or religious groups, in general groups with different beliefs and traditions, are always sure to be true.

What has been discussed above confirms that the comprehensibility of the world, so much emphasized by scientists and theologists, is always partial and Heraclitus was certainly right when, some millennia ago, wrote that "nature loves to hide," indicating that not everything in nature is understandable by the human reason. This fixed, we know from psychoanalysis that some deep aspects of the reality

---

[36]Both in the phylo- and in the onto-genetic infancy.

may sometimes be guessed by means of a knowledge modality different from the rational one, as it has apparently happened to some great personalities of science, of art and of spirituality of any epoch, as already discussed above.

In conclusion, if we look at things from the standpoint suggested by the most recent science, particularly the viewpoint showing the reality as a net of relations between us and the world, and not as a set of separate and independent objects, we realize that every branch of knowledge—science, philosophy, ideology, religion—is always subject to limitations and no paradigm is forever immunized against the knowledge error. In order to reduce to their minimum such limitations, it is necessary to always compare ourselves with others, with severity of judgment, freedom of prejudices, curiosity, and love for knowledge, without fanatisms or integralism.

3.2   Asserted and Denied Truths

It is established that we may not have always available the necessary means to ascertain the truth of an assertion related to an event or, especially, to a basic principle. However, there are historic or scientific truths, continuously and publicly discussed, on which we should not have any doubt, since they are always under the reflectors of our conscience, permanently submitted to accurate analyses and approved only after many years of experimentation and control. Unfortunately, within our societies, despite the continuous work of great social usefulness made by scientists, philosophers, and all other kind of scholars, fringes of enemies of the reason, lacking any independent and rigorous analysis' ability but blindly convinced of their own thoughts' autonomy, do not believe *a priori* to ascertained truths. This happens independently from any evidence is put in front of their attention, especially if suggested from their unconsciously accepted certainties, or if externally stimulated with evident scopes of enlistment to protest, when instead it would only need some level of concentration effort for a useful analysis of what really happens around us. Analysis that, for what previously observed (see chapter on rationality and common sense) is not always accessible to minds engaged in difficult social or psychological situations, or not accustomed to a constant mental application.

So, we find groups who, just for a blind defense of own prejudices, or sometimes due to lack of training to study and attention, or for personal social or political interests or, finally, just for the pleasure to contest, deny even irrefutable scientific truths. Two of the most surprising examples are related to the roundness of Earth and the Moon landing. Concerning Earth's shape, still nowadays we know about the existence of people stubborn to culture who contest its roundness, as if we were still in the medieval age, when the Copernicus' and Kepler's studies were accepted with mistrust together with the experimental observation by Tycho Brahe and Galileo Galilei (just to quote some of them), not talking about the further statements by other giants of science, like Simon de La Place, Albert Einstein, and many others.

Some of them still contest the landing on the Moon's surface of the Apollo 11 Space Mission with onboard the American astronauts Neil Armstrong and Buzz Aldrin, while the third spaceman Michel Collins was waiting in the lunar orbit within the Columbia command module, ready to bring back to Earth the spaceship. Great enterprise, this one, recorded and shown all over the world in all details, and notwithstanding it, contested till today by groups of uneducated rabid people, capable of refusing all what their insufficient knowledge does not allow to fully understand. For them, the *a priori* refusal remains the easiest available contesting possibility.

Not only science adventures, but also well-known social events are affected by refusal. There are, for example, those who deny the amply ascertained phenomena of enormous social importance like Hebrews' holocaust during the second world war, or those who denied for months the present COVID pandemic situation, despite the ascertained death of hundreds of millions of people all over the world. We could maybe agree that the denial attitude in this last case may escape from such a general analysis, and be justified by the instinctual angry reaction to the unforeseen turmoil that abruptly interrupted many human activities from which a lot of people draw their main sustenance.

Similarly interesting, I believe, is to mention the opposite cases of the denial theme discussed above, i.e., the one concerning the convinced and net acceptance of never ascertained truths, able to generate problems of similar seriousness. I think it, therefore, necessary to investigate such an item within the same discussion about presumed or ascertained truths—item that may also be indicated as 'claiming attitude' to mean the attitude by many people to give for certain something that has not been, or cannot be, proved as such.

Before starting this discussion, anyway, I find it important and necessary to extract from the assumed but not ascertained truths those basic principles to be accepted *a priori* because they are needed for the stability of the society itself, independent of any proof of their absolute truthfulness as, for example, in case of truths of religious type. The religious belief cannot be subject to experimental proofs, but the participation in such a feeling, measured by the enormous number of participating persons on Earth, does not allow to put it into discussion because it is based on the vital impulse of any human beings, mortal and fleeting on Earth, to believe in an extraterrestrial existence able to manage our future after the unavoidable death.

Anyway, even accepted as respectable the belief in God, we cannot hide the existence in the world of so many different, often even contrasting, religious beliefs resulting in a great variety of modes to conceive the divinity among the various human cultures, other than extra-human qualities, like omniscience and omnipotence. A variety of views that materializes in front of us as soon as we compare the many ways, official or not, that describe the religious feeling around the world, through the great number of dogmatic and ritual differences—like those specific to the hopes during and after life—all strictly dependent on the millenary traditions of any single societies, therefore variable from people to people because related to different cultures. It is a variety that frustrates the presumption of major truthfulness of one religious faith with respect to others; truthfulness that very often along the human history have been mistaken for absolute certainty within every single society. A claim that has often been taken with fury and safe from opposite faiths, so to cause the triggering of bloody wars, harbinger of huge massacres of innocent peoples, among the bloodiest in the history, phenomena occurring still today in many regions of the world.[37]

One of the strongest beliefs connected with religion has been for centuries the creationism, i.e., the belief that humans, differently from all other living beings, have nothing to do with the evolution of life on Earth, having been directly created by God together with the rest of the universe. A belief for which all phenomena occurring everywhere on Earth are strongly connected with the God's will and His capability to affect the course of events. Such a belief is supported still nowadays by many followers, despite the fact that, following the cosmological studies initiated by the priest-scientist George Lemaitre and those in the biologic sector started by Charles Darwin, it

---

[37]A kind of fury called 'fundamentalism.'

became definitely clear that our world was not created from nothing few thousand years ago, but that life on Earth exists from four billion years, i.e., just since its formation, probably arrived from sidereal spaces, maybe still in elementary forms, and it has evolved from then till the present human forms. Such knowledge, however, has not yet been able to erase at all the idyllic idea that human being, further than having been created only few thousand years ago, is also exempt from following the laws of evolution of life, at variance with the rest of all other living species.

The tendency to raise to absolute truths typical human phenomena, however, cannot be referred only to the concept of God. It is a fact, for example, that the notable socio-economic theory formulated by Karl Marx and Friedrich Engels is used by many to raise communism to an undisputable axiom, independent of the society's evolution. This tells us that the belief of a great number of people to consider a contingent social phenomenon as absolute and independent from history, is all the same able to enter overwhelmingly in the category of faiths, fiercely felt in a religious way.

Communism was born and started at about the end of the nineteenth century, when the birth and developments of important technologic activities started requiring the help of large masses of workers using unpleasant and alienating assembly chains—circumstance that produced humiliating submission relations between employees and employers with long and exhausting work shifts and scant salaries. Such a set-up has fortunately gone through profound modifications over the past century, both for the advent of new machineries able to consistently reduce (making it more human) the charge of work and for the important cultural activity done at those time by labor unions who, today, in the competition with economic groups for the influencing power, are changing their views and their consequent social behavior.

Communism, although ferociously opponent to all religions along the full course of the last century, has often assumed itself the characteristics of a religious belief based on the promoted but wrong idea that humans are all equal, idea that has implicitly assumed the meaning that we all have equal needs, whatever it may mean. A belief that, perhaps due to its intrinsic inconsistence, has given rise along time to the continuous appearance and disappearance of many groups,[38] all of them self-defined 'communist' although with great differences of practical connotations and often in reciprocal ideological contrast both for the primacy concerning the 'purity' of their specific principles, and for the way to define a universally accepted ideal, representative of the need to modify our society in all its bases in order to give rise to a society of 'equals'. I believe that pretending to build a society of 'equals' is an unrealistic and impossible objective to realize[39] since we humans are not equal and do not have all equal needs, but, exactly on the contrary, we are all different from one another as stated by the fact that our DNA is different for each single person. DNA,[40] in fact, tells us that human physicality and tendencies, although similar, are different for each one of us, circumstance easily recognizable by looking at the diversities of experiences we continuously have during our life. As such, DNA is used for individual identification.

To have mistaken 'similarity' for 'equality' has created the premises, apt for people humiliated and raddled because

---

[38]*Let us look at the following large list of Italian communist groups which has been members of the Italian Parliament: Democrazia Proletaria, Lotta Continua, Movimento dei Comunisti Unitari, Partito Comunista, Partito Comunista d'Italia, Partito Comunista d'Italia (marxista-leninista), Partito Comunista dei Lavoratori, Partito Comunista Italiano, Partito dei Comunisti Italiani, Partito della Rifondazione Comunista – Sinistra Europea, Partito di Unità Proletaria per il Comunismo, Sinistra Critica, ecc.*

[39]Also dangerous, as demonstrated by the birth of powerful totalitarian regimes all over the world!

[40]The molecule of Desoxyribo Nucleic Acid, previously cited.

excluded from the availability of the necessary means for a decent living, to see in communism the axiomatic characteristics of a religion, although opposed to classic religions, thus implicitly favoring the birth, from the Russian revolution ahead, of entire nations proudly defining themselves as 'communist.'

The so-called 'cold war' of last century between Soviet Union and the liberal Western countries was the dangerous consequence of such a situation that is furthermore amplified nowadays because of the presence in the world of several national states where it is established a strict control on their peoples apt to maintain it at a low civil state, equal for all except for those who manage the power and the richness of the nation. A condition that today, due to the presence of nuclear armaments, is dangerously near the explosion of a dangerous war between the most important political regimes of the world, due to the Russian expansionism at expenses of Ukraine. It is desirable a virtuous political process able to manage with caution the natural differences between persons discussed above, such to ensure the vital needs of everybody, even different, avoiding unacceptable excesses, both of power and of poverty, whose consequences may only be supporter of accumulation of richness in some places against increase of poverty somewhere else.

Probably an utopic idea, the last, that might remain impossible to put into operation even within any democracy, when its supervisors must inevitably be elected by a people of 'not equals' but considered 'equals' (one head, one vote), where the opinion of the frailest and less aware—but much more numerous—is predominant with respect to the more aware minority, especially because of the unavoidable presence of economic organizations, able to take benefit from such differences, thus changing at their own advantage what should instead be a free democratic process.

Indisputable truth that discussed above, that does not deny the fact—much more important for the whole human community—that we have never seen at the horizon (and still we do not see) a consistent motivated class of social and political groups able to realistically think to an ecologic future for any people in the world. A future respectful of the limited resources available on Earth, instead of thinking to the continuity of their personal power, keen only to maintain the existing, archaic, social, and economic *status quo*. A continuity that is only able to portend future disasters, as it is starting to happen with the progressive enlargement of the ozone hole that for centuries has defended our Earth from the deadly solar radiations of high energy.

3.3 The Exophysical Point of View

Near the end of the 19th century, the existence of living beings as 'ordered' physical systems was seen by Friedrich Wilhelm Ostwald and others in contradiction with the Second Principle of Thermodynamics. The discussions about such problem drove to the need of discussing the existing opinion for which humans are considered as being able to look at the world from outside, therefore neglecting their mutual interaction with the rest of the world, a perspective today referred to as '*exophysical* point of view.' According to such prevailing way of thinking of those times, it was considered normal and lawful to neglect the passage of time, so being able—at least in principle—to predict future events together with all their causal connections. The Laplacian determinism and Einstein's view that "the flow of time is only a stubborn human illusion," cited in Section 2.3, are just some of the most important consequences of the *exophysical* perspective, a point of view that has been disproved by science because clearly in contrast with our today's experience: it cannot be disproved that humans, same as all other living beings, are systems living far from thermodynamic equilibrium, which means that they have the need to continuously exchange information and energy with the external ambient. Humans need to assume high level organized energy, to release low level one, and to exchange information with other living beings. An exchange that, just because of the very high level of the intercommunications today reached within human society, allows us to extend our knowledge of the various aspects of reality in an always more complete and exhaustive way. A condition for which man is not seen anymore as he would be external to the world, but strictly connected to it, a condition that is indicated with the term *endophysics*, contrary to *exophysics*.[41]

It has already been discussed about the opinion—arisen at the beginning of 1900—of having nothing more to discover, and of the advent of the quantum theories that abruptly annulled it. Moreover, there was—and there is still today by many qualified scientists all over the world—the hope about the existence of a *theory of everything*. Today, with the unavoidable man–ambient interaction in mind, such a theory that would implicitly include humans theorizing it does not have any meaning. Talking, for example, about the extremely small, the heavy interaction between man and the nuclei in the moment of their measurement, renders it impossible to mix, within a unique independent theory, our description of the trajectories of electrons around the nuclei together with us who formulate it because the result depends just on the interaction itself. The most we can have, is just what we already have, i.e., a set of different scientific theories, each valid within the limits defined by the existing mutual interactions. The *exophysical* point of view may certainly be accepted when talking of the description of stars and planets—for example, by means of the universal gravitation theory and of the Relativity

---

[41]Cfr. Otto E. Rossler, *Endophysics: The World as an Interface*, World Scientific Publishing, London, 1998.

Theory—because of the negligible interaction between man and the celestial objects.

All that expressed above, especially considering the incredibly advanced technology that today characterizes the intense and continuous relations among all components of our modern society, should eliminate any residual doubt about the impossible adoption of the *exophysical* point of view and, consequently, on a possible *theory of everything* depending just on such a point of view.

3.4   Intersubjectivity, Endophysics

Today, our view of the external world is made up of a mix of data on all the various aspects of the reality. Data that we try to interconnect together in order to have available a solid mental construct, able to globally and univocally interpret the knowledge that we continuously acquire. Along the development of human societies, the mental constructs of single individuals started to incorporate and harmonize all the aspects of the knowledge connected to the social coexistence (laws, traditions, common activities, ...), an activity that gave a very important contribution to the development of the intersubjectivity, able to create a common vision of the world where all its aspects are interconnected. Obviously, each individual view, other than considering the great quantity of information continuously received from outside, includes also data not always personally experimented but assumed as true by intuition or by belief (ideologic, religious, politic, ...); data that are anyway necessary for a more complete and satisfying analysis of the contingent social, physical, and psychical needs of every one of us. Over time, the continuous interaction among all the single elements of the society submits every single mental construct to continue modifications. This leads to the progressive reduction of both the differences between individual views and the quantity of indirect and unconfirmed assumptions, especially those able to cause personal damage to individuals.

In addition, together with the evolution of a common view, the continuous social confrontation among single individuals contributes to the consensual definition of enquiry's methodologies and intercommunication rules, verbal and not verbal. A process, this one, that has produced a continuous updating of the communication language and that, at a certain degree of the social evolution, has also originated all those knowledge's disciplines that, as every other product of human intelligence, are subject to evolution in view of the continuous normal exchange of information among single persons.

The idea of 'intersubjectivity,' anyway, is not new. Immanuel Kant, in his *Critics of Pure Reason*, suggested that

> the comprehension of the world should be referred to the physic and perceptual peculiarities of man anchored to his ambient rather than to a presumptuous and impossible cosmic point of view from where, free from any interaction with the ambient, pretended to arrive to any possible degree of knowledge.

Such a concept was taken up by Karl Popper along a discussion in which he ironically invited to invent a new Laplace's demon,[42] able to make predictions on its ambient while staying within it, in the obvious consideration that nobody may be able to predict the results of his own predictions. Popper concluded that scientists cannot be considered spirits without body outside the universe, thus bringing forward the concept of *endophysics*, opposite to that of *exophysics*.

I find it necessary to underline here that the advent of quantum theories at the start of the last century should have

---

[42]Karl Popper, *The Open Universe: An Argument for Indeterminism*, 1988.

soon revealed the problem. The collapse of the wave function following any measurements reveals an evident irreversibility in the evolution of a phenomenon, and the Heisenberg *Indetermination Principle* shows evidence of the interaction between the observer and the object observed. Perhaps at those times the *exophysical* hypothesis was so strongly rooted that everybody preferred to interpret the collapse of the wave function as a voluntary act of the observer within the evolution of the universe. An interpretation that, even derived by Bohr's opposition to determinism, continues being followed today, notwithstanding the evident incompatibility between man's presumed free will and the insisting search of a *theory of everything.*

The *exophysical* hypothesis remains today as a surviving inheritance of classic physics, but the change of perspective is not certainly a trivial one since it implies a deeply different way to look at both science and the evolution of the human thought concerning the way our theories grow in our minds. Based on the *exophysical* hypothesis, many scientists, still today, look at mathematics as an entity independent of man who discovered its potentialities, based on his necessities or curiosity.[43] From the *endophysical* point of view, such an assumption—as well as the attribution of absolute truths to any results of scientific research—must be taken with caution. As already discussed, the two terms 'subjective' and 'objective' do not represent always a real dichotomy. The first refers to any individual perceptions not necessarily shared with others, while the second very often refers to rules of investigation and communication shared within some social groups in the world. Therefore, for as much we may increase and adjourn our investigation methods and our communication techniques, no one of our present or future discover may ever be considered with certainty as 'absolute.' Any development processes of

---

[43]An emblematic example is *Shadows of the Mind*, 1994, by Roger Penrose.

human knowledge in terms of the cycle individual–society–individual connected with the characteristics of perception and of conscious awareness is typical of the *endophysical* approach for which the ability of man to reach a complete abstraction from his ambient has to be excluded. A consideration that may be useful in order to understand that living beings too derive from nature and then are part of it, particularly human beings with their senses, their conscience, and their perception of reality, so to use with extreme caution the concept of 'absolute.' At this purpose, let me quote Friedrich Dürrenmatt when he says that

> every theory should only be considered a working hypothesis able to adapt to man's needs, more than a revealed truth to which everyone of us should adapt.[44]

3.5   Life vs Ambient

From the knowledge acquired over time, we have learned that starting from simple elements, the very long process of matter self-organization has produced on Earth the appearance of living organisms in close contact with their ambient from which they receive energy and information and to which they pour waste products. A process that has been able to produce life in almost every kind of physical condition by exploiting all available resources, even very poor ones, helped by the powerful solar energy. Humans, among all other organisms, pursue the goal of their survival and stability also by means of a path of knowledge allowing them, with time and experience, to optimally exploit such resources for any daily needs. All this suggests that we cannot think of human beings as separated by their ambient from which they receive any kind of material

---

[44]Cfr. Friedrich Dürrenmatt, *Nachgedanken* (*Nel cuore del pianeta*, Marcos y Marcos, 2003).

information, including all the necessary physical resources for their lives. Resources that, being not completely usable for humans' vital processes, are partially modified into wastes and returned to the ambient in order to be partially recycled and used for vital processes of inferior level, always present in our ambient, as it happens by millennia on Earth.[45] All living systems are clear examples of the continuous exchange processes of energy and information with the ambient in which they are merged. An exchange which occurs at the highest levels of complexity and through which the living acquire *Negentropy*, so increasing their order and complexity at the detriment of the ambient that, instead, increases its disorder.[46] An increase of complexity—at least concerning the development of human life—that may not necessarily occur gradually along time, as we might believe: it might 'emerge' in a jerky way as attested by the Stephen Gould's research, conducted with courage and with

> great discomfort due to the Darwinian belief for which any witness that cannot be inserted into a gradual sequence should be attributed to imperfections of the fossil documentation.[47]

For a harmoniously balanced echo-system, such an exchange of energy and information with the ambient would not put the problem of wastes in particularly important terms. This would be the case if any wastes of a living specie, animals, or plants, would be aliment for other species or if they could be used for other scopes, for example, to be recycled

---

[45]The recycling of wastes and scraps confirms, more than any other argument, the need to talk in terms of *endophysics*, leaving to *exophysics* the important role had along the development of knowledge.

[46]*Negentropy* is the opposite of *Entropy*, a measure of order.

[47]Stephen Jay Gould, *Punctuated Equilibrium* (*L'equilibrio punteggiato*, Codice Edizioni, Torino, 2008).

and reused, so realizing products useful for life. In these cases, only wastes not anymore used could go to increase the total entropy.

Unfortunately, as we may easily observe, too many wastes are today in front of our eyes, even along the streets of our towns. Wastes which we are not able to dispose of in time by adapting our public services with the connected operations of artificial re-cycling, or that are not reusable by other species, so preventing the ecosystem to be harmoniously balanced. All this generates enormous problems that, especially in large urban agglomerates, are often due to problems related to the overwhelming predominance of human beings with respect to other forms of life, which occurred during the last twenty-thirty years. Such a situation implies a continuous pathologic environmental pollution with an impressive accumulation of undisposed wastes. Accumulation that continuously sums up to the constant decrease of the biologic diversity derived by the already cited increase of the human population, that becomes always more difficult to manage in a society not yet used to the today levels of dis-equilibrium between humans and nature.

Let me propose some possible explanations for the difficulty to artificially compensate what nature normally did in the past. Besides the already cited huge increase in the human population, a possible answer could be the fact that we consume too much, well beyond our real physical needs. Very often our needs have a strictly psychologic value, as derived from the constant (sometimes also irritating) invitation, coming from producers of goods of any nature, to consume for just economic reasons, since their life is strongly bound to the production and sale of any kind of goods. It is a continuous and incessant call that invites people to consume, everything and always, without thinking about their real utility or about the consequential build-up of wastes.

The second answer concerns the civic costumes, then the citizen's sensibility to understand that a correct waste disposal is not only a duty, but a convenience for all of us in order to live in a decent common ambient, able to limit the possible arising of epidemies. Such simple rules should in principle be obvious to everyone, both to those who do not see the town as their home and to those administrators who do not exploit seriously the service for which they are paid (sometimes too much). The last ones should be blamed and punished even more than all the others, since they are paid by the community to seriously manage public utilities.

The bad civic habit and the excess of consumption, anyway, are not simple fortuities. They strongly depend on the operating modes of our society, very often outstretched toward a social development almost exclusively characterized by economic values to pursue at any rate. A condition, this, brought into being and continuously stimulated by sellers of everything in order to force an insane run to take over any kind of consumables, very often useless, especially in countries highly populated but where the consumes are still low. Such a way of doing, other than obscuring the ability of many people to fully understand the real needs of human life, impoverishes little by little the natural resources of our planet, destroys the biologic diversity and determines lack of resources, accumulation of wastes, and social disorders.

The indisputable truth discussed above does not deny the fact—much more important for the whole human community—that for a long time we have not seen at the horizon the rise of a consistent and motivated class of social and political groups able to realistically and factually think of an ecologic future, respectful of the limited resources available on Earth, instead of thinking of the continuity of their personal power, keen only to maintain the existing, archaic, social and economic *status quo.* A continuity that is only able to portend future disasters, as it is starting to happen with the progressive enlargement of the ozone hole that for centuries has defended our Earth from the deadly solar radiations of high energy.

Epilogue

Prof. Dr. Evgeni B. Starikov

The above essay demonstrates that Prof. Buccheri is absolutely right, up to one minuscule, yet truly peculiar point, which we would like to stress here.

Francesco La Mantia has caught this point in his *Afterword* to Prof. Buccheri's first essay, namely in paragraph '*6. Conclusions: Dialectics of Emergence*,' Page 168.[*] Prof. La Mantia speaks of '*construing the dismissing of classical objectivity*,' which ought to be a trend he is estimating as something throughout positive and worthwhile following.

In effect, this means that we do recognize availability of some kind of '*Emergence Dialectics*' but refuse to tackle the actual mechanistic details of the latter, by expressing our satisfaction with the fact that '*The non-equilibrium thermodynamics refers to non-isolated systems, open to interaction with outside. For such studies, the Nobel Prize for Chemistry was awarded to Prigogine in 1977*' (footnote 3, page 30, in Rosolino Buccheri, *Myth, Chaos, and Certainty: Notes on Cosmos, Life, and Knowledge*, Jenny Stanford Singapore, 2021).

---

[*]Rosolino Buccheri, *Myth, Chaos, and Certainty: Notes on Cosmos, Life, and Knowledge*, Jenny Stanford Singapore, 2021.

From the Top of the Mountain
Rosolino Buccheri
Copyright © 2025 Jenny Stanford Publishing Pte. Ltd.
ISBN 978-981-5129-22-9 (Paperback), 978-1-003-50829-8 (eBook)
www.jennystanford.com

Indeed, in his essay at hand, Prof. Buccheri concludes (*the present writer has dared to underline some parts of the text to follow, which are of relevance to his mind*):

'*My viewpoint avoids, as much as possible, to fall into the man's presumption to be the only intelligent being within the immense universe, anyway destined to perish together with it. Hard-to-die arrogance, this, perhaps derived by the many constraints,[†] both theoretical and practical, put in place by the numerous beliefs and religions in the world, that due to the impossibility to look into details within the Cosmos, are bound to the idea that life is special and unique on our Earth, and that it is bound to disappear together with Earth itself and with anything else here included*'.

This clearly reveals the traces of more than 100-years-old phantasm of the '*Heat or Chill Death of Universe*,' a result of Rudolf Clausius maxim:

[†]Of which '*constraints*' exactly Prof. Buccheri is speaking remains rather unclear, but as the entire story is dealing with methodological modalities, his epigraph cites Cartesian '*Discours de la méthode*,' and the story concerns the problems of the '*Second Law*' and '*Entropy*'—it is now logical to assume that it is just a reminiscence of Cartesian phrase, "*les actions de la vie ne souffrant souvent aucun délai, c'est une vérité très certaine que, lorsqu'il n'est pas en notre pouvoir de discerner les plus vraies opinions, nous devons suivre les plus probables*," and this is most probably to stress—and methodologically justify—some '*basic time constraints*' not letting us go in for actual details of the '*Second Law*' and '*Entropy*' problems, but, instead, forcing us to leave in full peace with the probabilistic entropy picture, and the Copenhagen interpretation of Quantum Mechanics... Noteworthy, there are absolutely no '*basic time-*' and/or '*geographic constraints*' for '*construing the dismissing of classical objectivity*'...

With this in mind, it is for the interested readership to decide, **which** kind of methodology we are dealing with here.

<u>The Energy of Universe is Constant,</u>
<u>The Entropy of the Universe tends to its Maximum</u>

Prof. Buccheri, likewise Prof. La Mantia, happens to be conceptually detained within the above framework, otherwise he would not ask: '*Entropy: a law or just a principle?*' (The correct answer to this poser: <u>*Neither... nor...!*</u>), and, thereafter: '*Is the Analysis Complete?*' The correct answer to the latter poser:

<u>*Yes, it has been complete more than 100 years ago, but you could not find the description of the relevant results in neither of Prof. Buccheri's essays, to our sincere regret. The actual point is that Rudolf Clausius has introduced the Entropy notion after thoroughly analyzing logical inferences of Nicolas Carnot and their pictorial representation by Benoît Clapeyron. Clausius has revealed Entropy's mathematical properties (its ever tending to its maximum value). But neither Clausius, nor the most of his contemporaneous colleagues and followers could manage analyzing—and finally—revealing the actual sense of the Entropy notion...*</u>

<u>*Hence, there is nothing more to 'construe' here.*</u>
<u>*Why?*</u>

Because '*construing the dismissing of classical objectivity*' has in effect nothing to do with the true processes of '*looking into details*' expected from righteous researchers and being surely non-trivial '*within the Cosmos*'—but is actually nothing more and nothing less than just a matter of personal choice for any researcher upon Earth, instead:

...To try plausibly justifying insistent ignorance—or—To try going in for details...

By the way, our Earth would never experience any kind of *'heat or chill death'*.
Why?

Entropy notion has absolutely no physical/chemical/biological, etc., sense, as soon as we try to apply it just as it is to any realistic system including the universe.

Entropy notion not only has sense but also is unavoidable, if we wish to explore the actual details of a *Realistic Process* in a *Realistic System* (of whatever nature).

Below, the present writer would like to present a concise sketch of the actual story. The interested readership might find all the necessary details here (cf. [1,2] and the references therein).

After attentively reading the deliberations, Prof. Buccheri provided us in his essay at hand, we recall that in general—and as a whole—the thermodynamics turns out to be a truly multilevel knowledge area. Still—at the first glance— the multilevel nature of the area seems to boil down to different fashions—flavors, tastes, smells—representations of thermodynamics having indeed practically nothing in common. To sum up, nowadays we deal with:

- *Equilibrium* vs. *non-equilibrium* thermodynamics;
- *Chemical* vs. *Engineering* vs. *Physical* thermodynamics;

Moreover, *'Entropy'* occurs to be much more than just *'the very Root of the universal Vis Viva,'* it is ostensibly *'Degree of Disorder,'* *'Arrow of Time,'* *'Measure of Information'*—and further unexpected examples of so-to-speak 「悟り」 (**'satori'**) forcing the basic, fundamental *'Entropy'* notion to teeter on the brink of the very semantic abyss.

Is it possible to somehow avoid the latter?

A Different Thermodynamics Is Carnot's Energetics.
May This Be Helpful?

Well, but it simply looks like answering a poser by posing a further question, sorry...

So, we don't seem to be capable of effectively avoiding the abyss just noticed in a flash...

The point is that we <u>do still have no clear idea</u> as to **what Entropy ought to be in fact**...

Likewise, we know that the old-and-good Quantum Mechanics—together with its entire realm—does **work very well** using **implicit** Entropy picture... What should then be our next revolutionary step? How should we put across the actual Carnot's Energetics?

Should we immediately launch a general worldwide hunt for, or even terror against the proponents of *Equilibrium* and *Non-Equilibrium* Thermodynamics?

<u>Definitely **NOT**</u>: Unlike what the Natural-Science-Revolutionaries have systematically compelled from their opponents—how the Natural-Science-Revolutionaries adulterated their opponents' lives, we shall instead duly hear to our opponents—just to try finding a mutually plausible solution to the common problem.

<u>OK</u>. If we now come back to Newton's mechanics, to the Newton/Leibniz controversy clarified by Roger Boscovich and Jean d'Alembert, and duly summarized, e.g., by Lazare Carnot, we do recall that <u>Potential Energy</u>, *Vis Mortua*, of any realistic system does come from the <u>interaction/coupling/correlation</u> among the relevant subsystems.

Vis Viva (Kinetic Energy) comes from *Vis Mortua*; it enlivens/produces <u>Actions</u> and, consequently, encounters all the ubiquitous <u>Reactions/Counteractions</u>: various, diverse, innumerable, uncountable hindrances, interferences, impediments, baulks, checks, deterrents, handicaps, encumbrances, hitches, preventatives/preventives, obstacles, obstructions, impedances, resistances, resistivity, etc., ...

Whatever *Actions/Counteractions/Reactions* ought to be, they should comply with Hon. Isaac Newton's *Third Basic Law of Mechanics*:

Zero Action means Zero Reaction;
The more Action, the more Reaction;

Attentive readership would exclaim here: so, the Counteractions/Reactions might well **go to Infinity this way**, aren't they?

Good news (by Nicky Carnot, Rudolf Clausius, etc.): **Reaction must always reach its Maximum.**

May Action have intensity enough to *compensate/equilibrate* the Reaction Maximum—the Process would reach its Ultimate Aim.

The above is just what we may dub: **Enthalpy-Entropy Compensation/Equilibration**, if we would recognize the **Heat Content (Enthalpy)** as the universal source of **Actions**—while uniting the ubiquitous, various, diverse **Reactions/ Counteractions**: innumerable preventatives/ preventives, hindrances, hitches, obstacles, encumbrances, impedances, interferences, impediments, deterrents, handicaps, obstructions, resistances, resistivity, baulks, checks, etc., under the common term of **Entropy**.

Indeed, it is the famous Free (Available) Energy formula by J. W. Gibbs:

$$\Delta G = U + PV - TS,$$

i.e.,

$$\Delta G = H - TS,$$

where

U is the internal energy (SI unit: Joule),

P is pressure (SI unit: Pascal),

V is volume (SI unit: m³),

T is the (absolute) temperature (SI unit: Kelvin),

S is the Entropy (SI unit: Joule per Kelvin),

H is the Enthalpy (SI unit: Joule).

Gibbs formula does render **implicit** the interplay among ALL the relevant Actions (the Enthalpic term) and ALL the pertinent Counteractions (the Entropic term).

Noteworthy, we see that *liquid viscosity and friction of any kind* do really belong to **entropic effects**—whereas Gibbs formula does inevitably bring us back to

* *Equilibrium* and *non-equilibrium* thermodynamics;
* *Chemical/engineering* and *physical* thermodynamics;
* ...

The attentive readership would immediately recognize that the conventional stances are not that bad. The basic and fundamental **Entropy** notion hasn't disappeared anywhere—All the conventional thermodynamic branches are taking it into account **properly** but **implicitly**. Hence, the inevitable basic posers arise:

Is it OK to treat Enthalpic and Entropic terms **explicitly**? If so, **what may be the benefits**?

The standard approach suggested by the conventional thermodynamics is OK for the **implicit** Enthalpy-Entropy picture; however, employing it, e.g., for studying *chemical reaction mechanism's* **fine details** is somehow likewise *eating soup with a fork*... Indeed, simulations of complicated systems at the explicit full-atomic level—that is, using any progressive modern QM/MM combination—would not guarantee a breakthrough without the explicit analysis of the **microscopic Action/Counteraction** interplay.

The above may clarify the present writer's striving for common collegial efforts: Indeed, not for adulterating each other's lives, as ever, but for duly trying to find mutually plausible solutions to the common problems.

Second Law of Thermodynamics

As we speak of Entropy, we also need to touch the '*Second Law of Thermodynamics.*'

We are sincerely persuaded that the '*Second Law of Thermodynamics*' must be a fundamental law of nature, a one of the most valuable discoveries of mankind.

Howbeit, this law is steadily confusing for most engineers or students. The main reason for this is that it contains so many 'complex terms' and there are so many ways to phrase this law, but—most importantly—the majority does not even have the foggiest what the actual applications of this law are.

Even big scientific research workers' brains had and still have to stumble herewith...

Usual reports on the topic, e.g., the essay at hand, do demonstrate the meanwhile widespread *Operational Positivism* (**this is an unwillingness even to touch, not to speak of going in for, important details**) the readership might easily muddle up with the '*Engineering Rigor*'...

Hence, a kind of clarification ought to be urgently necessary! So, captain, AHOY!

A. **There should be ONLY ONE BASIC, fundamental Energy Conservation and Transformation Law**. It is definitely unique—in delivering **two** logically **joint concepts**, which are **conceptually indivisible**—these are *Energy Conservation*[‡]—and *Energy Transformation*.[§] Still, a more-than-100-years-old conceptual failure has brought us to two separate thermodynamic laws—but this has nothing in common with the actual physics. To come back, they have coined two more fake thermodynamic laws, employed the *Probability Theory + Mathematical Statistics*, and this has helped formulate the *Quantum Mechanics*, which did become a *Valid and Seminal Physical Theory*, a worthy result of immense international efforts, in remaining a *basically metaphysical conceptual construction* (for details, see [1,2] and the references therein)—so, *Quantum Mechanics*, **methodologically**, ought to be only restrictedly fruitful.

B. By dividing the basically indivisible law, we are touching Combinatorics, we are touching Probability Theory, we are even stepping back to Thermodynamics for a while, but...

We are still NOT correctly answering the poser: WHAT IS ENTROPY? Sorry!

1. In the formula $S = k_B \times \ln(\Omega)$, we do imply, Ω means not a '*Huge Number of Microstates*,' and not '*Probability*' of God knows what *scilicet*, which numerically ranges between [0,1] by the way, not even '*Wavefunction*,' which ought to be just a purely metaphysical notion, as it is... In effect, Ω ought to be just a simplistic algebraic function of Lord Kelvin's Absolute Temperature. Dr. George Augustus Linhart has published this result 100 years ago in *JACS*. By the way, Prof. Gibbs had learned **What ENTROPY Exactly IS** while attending chemical thermodynamics lectures by Prof. Horstmann in Heidelberg, whereas Prof. Horstmann was one of the

---

[‡]**Quantitative** feature of the **Energy** notion.
[§]**Qualitative** feature of **this same (!!!) Energy** notion.

very students of Prof. Clausius. Prof. Gibbs was duly heading for the proper embodiment of the Clausius–Horstmann ideas (in fact, Carnot–Clausius–Horstmann ideas, to be fully correct!), but he had not enough time for accomplishing this, to our sincere regret... Dr. George Augustus Linhart was one of the <u>very rare true followers of Prof. Gibbs</u>, but he had a very bad luck...

2. ...It's a truly long, interesting and instructive story (for details, please cf. [1,2] and the references therein)...

3. WHAT-ENTROPY-IS-poser has been answered not by Clausius, not by Boltzmann, etc., but by Goethe, who has introduced Mephistopheles, the perfect philosophical embodiment of the ENTROPY notion.

4. Newton did basically know WHAT ENTROPY IS—<u>A Counteraction</u>.

5. That <u>Counteractions</u> do never grow to *Infinity* with the growing <u>Actions</u>, but **MUST** reach their **MAXIMUM** values, is the result by Nicky Carnot, which has been formalized by Rudolf Clausius...

Thus, the main poser, **not answered** by the conventional *'Equilibrium' Statistical Thermodynamics* ought to be:

WHAT IS ENTROPY???

It is but **Carnot's Energetics** that suggests **the proper answer**—based upon the classical Newtonian Mechanics.

For our Chinese colleagues this ought to be (太極圖, tàijítú), representing (太极; tàijí; 'utmost extreme') in both its monist (無極, wújí; literally: 'without ridgepole,' meaning 'without limit') and its dualist (陰陽, yînyáng, just a Chinese philosophical concept describing opposite but interconnected forces) aspects. In the academic and engineering parlance we may speak here of the basic, fundamental *Action-Counteraction couplings* obeying Hon. Isaac Newton's Third Basic Law of Mechanics.

Furthermore, for our international colleagues, this ought to be **Mephistopheles**, who represents the brightest possible '*personalized*' embodiment to the 太極圖:

Faust: "*Nun gut, wer bist du denn?*"

Mephistopheles: "*Ein Teil von jener Kraft,*

Die stets das Böse will und stets das Gute schafft."

[*Johann Wolfgang von Goethe*]

Faust: "*Howbeit, who are you then?*"

Mephistopheles: "*A Part of Power, which would*

The Evil ever do, and ever does the Good."

[*Transl. by G. M. Priest, taken from Internet*]

This is how to start discussing the poser: "*What the sense of ENTROPY notion in fact is.*"

Therefore, the sense of the 太極圖 in the academic and engineering parlance ought to be the **Enthalpy-Entropy Compensation/Equilibration** notion:

Isaac Newton: Zero Action, Zero Reaction; The more Action, the more Reaction;

Good news (by Carnot, Clausius, etc.): Reaction must always reach its Maximum.

Does Action compensate/equilibrate Reaction—the Process does reach its Aim.

This is how to start discussing the poser: "*What the Second Law of Thermodynamics in fact is.*"

But how general is this phenomenon at all?

The Core Phenomenon of the '*Second Law*' ought to be long known to our Chinese colleagues *via* the 易經 (*I Ching/Yi Jing*)—usually translated as the '*Book of Changes*' or '*Classic of Changes.*'

Here, 易 = **easy**, 經 = **through**, i.e., 易經 = "A recipes collection as for how to Come Through with Ease." The main philosophy behind this ought to be (無極, wújí, i.e., '*Without Limits*'), i.e., the 易經 should be a recipes collection for ensuring how to always come through with ease, irrespective of any kind of limitations/interferences.

To sum up, the above is about the '**Entropy**' in Physics, Chemistry, Biology, etc., etc., etc.

Entropy in other fields:

Beware of a powerful trend to build up misnomers in many other fields, for *Entropy* is the term **solely** for the 'Energy Transformation,' whatever nature/origin this energy and its transformation might be of.

The Arrow of Time:

This is **not due solely** to Entropy, it is owing to the *Eternal Action-Reaction Interplay, i.e., Entropy-Enthalpy Compensation.* This is why the '*Universal Heat/Chill Death*' morons are stubbornly proclaiming ought to be just a useless historical legacy.

The origins of randomness:

Howbeit, the actual **origins** of randomness **lie solely** in our **ignorance** and/or in our *unwillingness/incapability to lift the latter.* Hence, there might be absolutely no other explanations to furthering some '*deepest intrinsic senses of randomness*' than plainly the marketers' strive for demonstrating their highest skills in mathematically dealing with ubiquitous random processes.

✑✒

The overall impression both essays by Prof. Buccheri have left causes to revert to immortal '*Divina Commedia*' by Dante Alighieri, namely to the third Song of The Hell. The present writer has decided to cite the original and to post two own translations, the one in Russian, his mother tongue, and the one in English.

Finally, the present writer wishes to duly conclude his epilogue by sharing with the readership his humble opinion as to the current state of art in the field.

Canto terzo

Per me si va nella città dolente;

Per me si va nell'eterno dolore;

Per me si va tra la perduta gente;

Giustizia mosse il mio alto Fattore;

Fece mi la divina Potestate,

La somma sapienza, e il primo amore.

*Dinanzi a me non **far** cose create,*

Se non eterne, ed io eterno duro[¶];

Lasciate ogni speranza, o voi che entrate.

Queste parole di colore oscuro

Vid'io scritte al sommo d'una porta;

Per ch'io: Maestro, il senso lor m'è duro...

༄༅

[¶]These both lines in the original Italian editions (starting from that of 1830, at least) *do seem to contain a typo*: '*Dinanzi a me non **fur** cose create, Se non eterne, ed io eterno duro*' do read we here, which ought to dictate the following straightforward orthographical amendment: '*Dinanzi a me non **fuor** cose create, Se non eterne, ed io eterno duro*' and, consequently, entails the English translation: '*Before me there were no created things, If not eternal, though I do last forever*'. Still, this is well known to extremely complicate the interpretation. As Dante describes herewith the Legend on the Top of the Gates to the Hell, we do feel it most appropriate to suggest here *far* instead of *fur*, which ought to enable reasonable comprehension and – consequently – proper translations. It is definitely of interest to learn from the specialist who does have the access to the very original manuscript, whether our suggestion here might really work out...

Третья песнь

Через меня войдёшь ты в Город Боли;

Через меня познаешь Вечную ты Боль;

Через меня придёшь к заблудшим ты;

Мой Высший Движитель суть Справедливость;

А сотворён же я Божественною Силой, и

Высшей Мудростью, и Первою Любовью.

Передо мной да не твори,

Коли не вечен ты, ибо на то здесь вечен Я;

Оставь надежду, всяк сюда входящий.

Да, эти мрачные слова увидел я

Как надпись на Перекладине Ворот;

На что я: Мастер! О, как же трудно смысл их постичь мне...

ை

Third song

Through me Thou enter Pain City,

Through me Thou learn Eternal Pain,

Through me Thou come to those gone astray.

As my Supreme Contractor ought to be The Justice,

Divine Power and The Supreme Wisdom,

The Very Love ought to be my Movers.

Were thou not eternal, don't act and/or perform in front of me,

For I, instead, do last forever...

Abandon all thy hopes, ye who enter here.

Oh, yes, it is just the gloomy words that I have noticed,

As written on the Top of Gates;

To which I: Oh, Master! Their meaning is so hard for me to comprehend...

...Coming back to the topic of our discussion, the Entropy notion, aside from all the fluffy misnomers surrounding it, does essentially belong to the fundamental ones truly so hard for us to comprehend.

Howbeit, the numerous colleagues who have done their very best to duly but IMPLICITLY incorporate the Entropy notion into physical/chemical theories have definitely not gone astray. The present writer is not meaning Nobel Prizes here.

The stories of the colleagues who have gone their own ways to duly clarify the Entropy notion and finally shed light upon the possibilities of using it in our theories EXPLICITLY had much more curvilinear *curricula vitae*, as compared to the above group.

But, fortunately, <u>manuscripts are not deflagrable.</u>** Now, it is up to the readership to decide, whether this or that approach, or even their skillful combination(s) may truly help us pursue our scientific research work.

Dr. Evgeni B. Starikov
Graduate School of System Informatics
Kobe University
1-1 Rokkodai, Nada, Kobe 657-8501, Japan

Chemistry and Chemical Engineering
Chalmers University of Technology
Göteborg, Sweden

** "*Рукописи не горят*" (М. А. Булгаков, "Мастер и Маргарита").

If science is in principle an antidote against theology, it is also true that the last ends up by impressing its brand even in science, ends up by infecting it. Theology has injected into science the syndrome of the 'will of truth'. The same that has already infested philosophy... The will of truth has put always together the two sisters, philosophy, and theology, vainly stretched to the search of a stable, absolute, truth, able to bear the collision with the world's chaos and irrationality. The will of truth is synonymous of fragility. It means that the truths cannot be grasped. The will of truth is will of dominance since it aims to possess the truth.

From DIE FRÖHLICHE WISSENSHAFT,
by Friedrich Nietzsche

Bibliography

Mario Alai (ed.) *Il realismo scientifico di Evandro Agazzi*, Atti del convegno di studi di Urbino del 17 Novembre 2006, Montefeltro, Urbino 2009 (edizione speciale di *bonomia*).

Marina Alfano and Rosolino Buccheri, *Oltre la razionalità scientifica*, Lateranum, pp. 245–265, 2012/2.

Jim Al-Khalili, *Il mondo secondo la fisica*, Boringhieri, 2020.

John Barrow, *The Constants of Nature. From Alpha to Omega*, Jonathan Cape, 2002.

Valentin Boriakov, Rosolino Buccheri, and Franco Fauci, *Discovery of a 6.1 ms binary pulsar PSR1953+29, Nature*, 304, 417–419, 1983.

Rosolino Buccheri, *Fra il mito della certezza e la certezza del mito. L'evoluzione della conoscenza fra legge e casualità*, Carlo Saladino, 2019.

Rosolino Buccheri, *Myth, Chaos, and Certainty. Notes on Cosmos, Life, and Knowledge*, Jenny Stanford Publishing Pte. Ltd., 2021.

Rosolino Buccheri, George Jaroszkiewicz, and Metod Saniga, *Endophysics, the fabric of time and the self-evolving universe, Research Signpost*, 609–623, 2003.

Marta Burgay et al., *An increased estimate of the merger rate of double neutron stars from observations of a highly relativistic system, Nature*, 426, 531–533, 2003.

Fritjof Capra, *La rete della vita. Una nuova visione della natura e della scienza*, Edizione CDE spa—Milano, 1997.

Marco Tullio Cicerone, *Somnium Scipionis*, Ciranna, 2003.

Antonio R. Damasio, *L'errore di Cartesio. Emozione, ragione e cervello umano*, (*Descarte's Error. Emotion, Reason, and the Human Brain*), Adelphi, 2008.

René Descartes, *Discours de la méthode* (*Discorso sul metodo*), Milano, Mursia, 1972.

Friedrich Dürrenmatt, *Nachgedanken*, Diogenes Verlach AG, Zurich, 1998.

Arthur Stanley Eddington, *La natura del mondo fisico*, 1915.

Albert Einstein, Boris Podolsky, and Nathan Rosen, *Can quantum mechanical description of physical reality be considered complete?*, Physical Review, vol. 47, p. 777, 1935.

Nils Engelbrektsson (1875–1963), Karl Franzén (1882–1967), Evgeni B. Starikov, *The basic features of thermodynamics*, Monatshefte für Chemie—Chemical Monthly, v. 152, pp. 1437–1490, 2021.

Marcelo Gleiser, *A Tear at the Age of Creation: A Radical New Vision for Life in an Imperfect Universe*, Free Press, 2010.

Elkhonon Goldberg, *Il paradosso della saggezza. Come la mente diventa più forte quando il cervello invecchia* (*The wisdom paradox. How your mind can grow stronger as your brain grows older*), Ponte alle Grazie, 2005, Milano.

Margherita Hack, Pippo Battaglia, and Rosolino Buccheri, *L'idea del Tempo*, UTET, 2005.

Stephen Hawking, *The Theory of Everything. The Origin and Fate of the Universe*, New Millennium Press, 2002.

Stephen Hawking and Roger Penrose, *The Nature of Space and Time*, 1996, Princeton University Press.

Michał Heller, *Creative Tension: Essays on Science and Religion*, Templeton Foundation Press, Pennsylvania, 2003.

James Hillmann, *On paranoia & on the necessity of abnormal Psychology: Ananke and Athena*, Eranos Jahrbuch, LIV, 1985 & XLIII, 1974.

Russel Alan Hulse and Joseph Hooton Taylor, *Discovery of a pulsar in a binary system*, NASA Ads, 1974.

Immanuel Kant, *Critica della Ragion pura*, 1781.

Johannes Kepler, *Somnium. Opera postuma sull'astronomia lunare*, Edizioni Theoria, 1984.

Thomas Kuhn, *Dogma contro critica. Mondi possibili nella storia della scienza*, Raffaello Cortina, 2000.

Pierre Simon de Laplace, *Essai Philosophique sur les probabilites*, 1814.

James Lovelock, *Gaia. A New Look at Life on Earth*, Oxford University Press, 1979.

Jacques Monod, *Le hazard et la nécessité*, 1970.

Philip Morse, *Thermal Physics*, W.A. Benjamin, inc., New York-Amsterdam, 1964.

Ernest Nagel, *Gödel's Proof*, New York University Press, 1968.

Friedrich Nietzsche, *La gaia scienza*, LIBRITALIA, 1997.

Eldredge Niles, Stephen Jay Gould, *Punctuated equilibria: An alternative to Phyletic Gradualism*, in T.J.M. *Schopf Models in Paleobiology*; Freeman, Cooper & Co, San Francisco, 1972.

Piergiorgio Odifreddi, *La matematica del novecento*, Giulio Einaudi, Torino, 2000.

Roger Penrose, *Shadows of the Mind*, Vintage, 1994.

Karl Popper, *The Open Universe: An Argument for Indeterminism*, Routledge, Taylor & Francis, 1988.

Ilya Prigogine, *La fin des certitudes : Temps, chaos et les lois de la nature, Editions Odile Jacob, Paris, 1996*.

Ilia Prigogine and Isabelle Stengers, *La nuova alleanza. Metamorfosi della scienza*, Piccola Biblioteca Einaudi, 1999.

Jeremy Rifkin, *Entropy. Into the Greenhouse World*, Viking Penguin, 1989.

Otto E. Rossler, *Endophysics: The World as an Interface*, World Scientific Publishing, London, 1998.

Bertrand Russell, *Autorità e individuo. I doveri dello Stato e i diritti dei cittadini*, Longanesi & C, 1970, Milano.

Metod Saniga, *Algebraic geometry: a tool for resolving the enigma of time?* in *Studies on the Structure of Time. From Physics to Psycho(patho)logy*; Rosolino Buccheri, V. Di Gesù, and Metod Saniga, Kluwer Academic/Plenum Publishers, 2000.

Lee Smolin, *The Life of the Cosmos*, Oxford University Press, 1997.

Evgeni B. Starikov, *How many laws has thermodynamics? What is the sense of the entropy notion? Implications for molecular physical chemistry*, Monatshefte für Chemie—Chemical Monthly, 152, 871–879, 2021.

Guido Tonelli, *Genesi. Il grande racconto delle origini*, Universale Economica Feltrinelli, 2020.

Susie Vröbel, *Fractal Time. Why a Watched Kettle Never Boils*, World Scientific, 2011.

Name Index

Subject Index

Printed in the United States
by Baker & Taylor Publisher Services